THE ENIGMA OF
FERMENT

FROM THE PHILOSOPHER'S STONE TO
THE FIRST BIOCHEMICAL NOBEL PRIZE

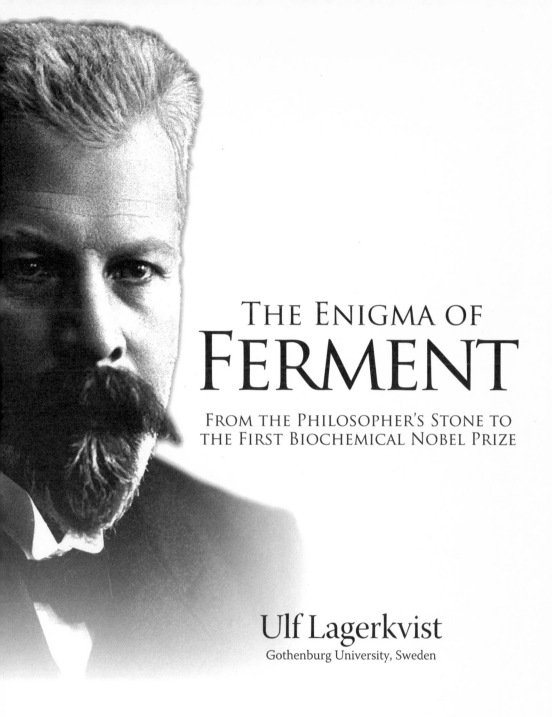

THE ENIGMA OF
FERMENT

FROM THE PHILOSOPHER'S STONE TO
THE FIRST BIOCHEMICAL NOBEL PRIZE

Ulf Lagerkvist
Gothenburg University, Sweden

World Scientific

NEW JERSEY · LONDON · SINGAPORE · BEIJING · SHANGHAI · HONG KONG · TAIPEI · CHENNAI

Published by

World Scientific Publishing Co. Pte. Ltd.

5 Toh Tuck Link, Singapore 596224

USA office: 27 Warren Street, Suite 401-402, Hackensack, NJ 07601

UK office: 57 Shelton Street, Covent Garden, London WC2H 9HE

British Library Cataloguing-in-Publication Data
A catalogue record for this book is available from the British Library.

THE ENIGMA OF FERMENT
From the Philosopher's Stone to the First Biochemical Nobel Prize

ISBN 981-256-421-7 (pbk)

Printed in Singapore by Mainland Press

Contents

Preface

The aim of this book is to give a popular account of the history of the ferment, that from its obscure origin in medieval medicine and alchemy, gradually developed into the modern concept of the enzyme. The history of ferment and generally of biochemistry has always been intimately associated with that of medicine. It is therefore necessary to begin the book with a fairly extensive introductory chapter that attempts to summarize the development of medicine from the days of Hippocrates to the scientific revolution in the 17th century. In the following chapters, the focus is on the transformation of chemistry and biochemistry from a collection of recipes to a science in its own right, with an intellectually satisfactory theoretical background comparable to that of physics, the most advanced of the natural sciences.

During the 19th century, the question of the nature of ferment leads to the long drawn out, bitter conflict between the "vitalists," who believed in a "vital force" peculiar to the living cell, and their opponents, the "chemists" who regarded "vital force" as superstition and instead advocated chemical explanations of the same kind as in ordinary test tube experiments. The dispute can be said to have centered on the question of alcoholic fermentation. Was this the exclusive privilege of the living yeast cell or could it in principle take place outside the cell, catalyzed by soluble ferments? For a long time the vitalists, under their chieftain Louis Pasteur, seemed to have the upper hand, but finally in 1897 Eduard Buchner was able to demonstrate cell-free alcoholic fermentation in an extract of yeast. This definitely put an end to vitalism and at the same time won Buchner a Nobel Prize, the first to be awarded for a purely biochemical work. The last part of the book is focused on this event.

Finally, it should be pointed out that this book is intended primarily for the layman reader and does not presuppose any knowledge of biochemistry and enzymes. At the same time, the author wants of course to present something more than entertaining chatter and hopes to strike a reasonable balance between scientific ambition and the wish to popularize. I am greatly indebted to the Royal Swedish Academy of Sciences and its Chemical Nobel Committee for generously giving me access to their archive with information on Eduard Buchner. I am very grateful to the librarian of the Academy of Sciences, Christer Wijkström, for valuable help with the illustrations of this book.

I am greatly indebted to Jacques Fresco who was kind enough to read the manuscript and give helpful opinions.

INTRODUCTION

Chemistry is the science that historically is closest to medicine. Even in its most primitive form, medicine has attempted to use extracts and potions from primarily the vegetable kingdom to create a kind of pharmacopoeia in order to treat diseases. It is easy to see that collecting herbs and subjecting them to different operations to obtain preparations with improved medicinal qualities could eventually lead to procedures that are close to what might be called chemistry. In what follows, we will briefly outline the history of medicine from the golden age of Greek culture to Harvey's discovery of the blood circulation.

Medicine in Ancient Greece

In the West, we have a tendency to see the Greek culture as the triumph of rational thinking and individualism. In the medical field this is certainly true of the scientifically oriented Hippocratic school, even if there was also another, religious-magical line in Greek medicine symbolized by the worship of Asclepius, the God of healing, and strongly influenced by traditions inherited from ancient Egypt.

Hippocrates, by far the most influential of all medical writers in ancient Greece, was born on the island of Cos, probably about 460 BC, and died at a high age around 370 in Thessaly. With these rather meagre biographical data we have on the whole exhausted all that we know with certainty about the father of medicine. We do not even know if the approximately 60 books traditionally ascribed to him, really have Hippocrates as their author. These writings, often referred to as *Corpus Hippocraticum*, are a rather heterogeneous collection of medical treatises from different time periods by a number of authors, compiled at the end of the 4th century BC and motivated by the needs of the great library in Alexandria. In all probability, Hippocrates is the author of only a few of all the books that have been ascribed to him as the most famous representative of Greek medicine.

The modern reader is impressed above all by the very exact clinical observations of Hippocrates and his detailed description of the patient's condition that often enables us to arrive at a diagnosis more that 2000 years later. Nevertheless, this is not the reason for the immense, all-embracing influence of Hippocratic medicine, from the 4th century BC to the middle of the 19th century. To understand its dominating position, we must turn to the theory about the four body humors. The Hippocratic physicians had inherited this concept from Egyptian medicine, but they developed it into an apparently scientific, all-explaining medical theory based on the idea that health and disease were dependent on the relations of the body humors to each other. The famous treatise *On Human Nature* contains the following summary of the theory: "The human body is made up of blood, phlegm,

Venesection.

Courtesy of Bonnierförlagen, Stockholm, Sweden.

yellow bile and black bile, and this is what constitutes the bodily nature of man and determines sickness and health. Therefore, man is healthiest when these fluids are properly mixed and well proportioned with respect to each other in terms of strength and quantity. He suffers, on the other hand, when any of these components is present in a quantity that is too small or too large, or when it is sequestered in the body and not mixed with the other humors." It should be pointed out that even if in theory the four humors were equally important in Hippocratic pathology, it was in reality the phlegm that dominated medical thinking, and both the prophylactic and therapeutic measures of the physician were aimed at keeping the dangerous phlegm in check.

The impact of the theory of the four body humors was greatly enhanced by the early coupling of it to the ancient doctrine of the four elements and their well-known properties: Air (warm and moist) = the blood; water (cold and moist) = the phlegm; fire (warm and dry) = the yellow bile; earth (cold and dry) = the black bile. From this relationship there arose in the Middle Ages the popular idea of the four temperaments. A sanguine temperament, when you are happy and optimistic, is caused by an excess of the warm blood (sanguis). The melancholy temperament, on the other hand, can be explained by an excess of black bile (melancholia); hot-headed and choleric individuals have much yellow bile (cholera), while the cold phlegm dominates in the calm and thoughtful. It is not to be wondered at that this simple and straightforward, all-explaining theory became immensely popular and had an enormous influence on medical thinking.

The Hippocratic physicians were not so concerned with a correct diagnosis, which we today consider to be the very foundation of medicine. However, their somewhat cavalier attitude becomes more understandable in view of the Hippocratic theory that all diseases are caused by disturbances of the body fluids and their relative proportions. Regardless of the diagnosis, the therapy would always be aimed at restoring the correct balance of the body humors. On the other hand, if they were not so interested in distinguishing one illness from another, the prognosis (the ability to predict the outcome of an illness) was the more important to the Hippocratic physicians and largely determined their reputation as doctors. No doubt, their power of observation at the sickbed, so much admired over the ages, was a decisive factor here.

A considerable part of *Corpus Hippocraticum* is devoted to surgery, and this subject is dealt with in a refreshingly straightforward way, free of speculations and unwarranted hypotheses. Of all that Hippocrates supposedly wrote, this comes closest to our modern view of medicine. His powers of observation are truly amazing and his clinical methods might in many cases, for instance the reducing of dislocations, be used even today. The Hippocratic writings demonstrate a detailed and accurate knowledge of the anatomy of the skeleton, and to some extent also of the tendons and ligaments, no doubt the results of an extensive surgical experience. His knowledge of the internal organs is much less precise. As in other ancient cultures, systematic dissections were not possible in Greece for religious reasons.

For all diseases that fell within the province of what we today call internal medicine, the Hippocratic therapy consisted of a rather strict dietary regime and of course various kinds of bloodletting. The drawing of blood by venesection and cupping, i.e. scarification of the skin followed by different methods of suction, has been associated with classical medicine for thousands of years. The reason that it became so immensely popular and remained the incomparable therapeutic method for so long, was probably because in certain cases, the best known example being pneumonia, it gave the patient a feeling of relief from his breathing difficulties even though the bloodletting in reality worsened the prognosis.

The rationale behind bloodletting was obviously that in this way one could directly attack the cause of the disease by removing pathologically altered blood, and in particular the dangerous phlegm, from the body. As far as cupping is concerned, it is easy to see that this procedure might often be thought to evacuate the sickly blood specifically from the affected part of the body. However, venesection also permitted an effect that was selective and directed against the diseased organ. Curiously enough it was believed that for each organ there was a certain corresponding vein. Thus, in the case of pneumonia a particular cubital vein was opened, the exact choice of vein depending on which lung was thought to be diseased. Clearly, the ancient Greeks realized that the blood could move within the vascular system, but the idea of a blood circulation was alien to them.

Even if the *Corpus Hippocraticum* is mainly concerned with clinical problems, Hippocratic medicine certainly has a theoretical foundation. It is highly speculative and contains a number of assumptions and ideas that in our eyes are fantastic, not to say unreasonable. One may then ask for instance what it is in the doctrine of the four body humors that is worth admiring? The answer is of course that even if the Hippocratic therapy in the case of internal diseases was probably no more effective than the methods practiced in Babylonian or Egyptian medicine, there is nevertheless a great and decisive difference between Hippocrates and his ancient predecessors. Hippocratic medicine stands entirely on a scientific basis, however fanciful and inaccurate; all ideas about gods and demons as the direct cause of sickness and health have been thrown into the dustbin of medical history. Regardless of how strange many of the Hippocratic theories may seem to us, getting rid of the magic element in medical thinking was a major step forward. That is the reason why modern medicine may be said to have its roots in the golden age of Greek culture.

Classic Medicine After Hippocrates

The Alexandrian School

In our days, anatomy is hardly the spearhead of medical research. It is more than a hundred years since any really important discoveries were made in this field. Dissections of human bodies play absolutely no role in scientific research today, but this has not always been so. At the peak of Alexandrian medicine in the 3rd century BC, a better understanding of anatomy was generally recognized as being of crucial importance for any medical progress. Obviously, systematic dissections of human bodies were an absolute prerequisite here. It was therefore a great breakthrough for medical science when the Ptolemaic rulers in Alexandria, where Greek medicine now had its center, for the first time in antiquity allowed physicians to study human anatomy without any religious restraints. It is true that this was possible only for a comparatively short period during their enlightened rule; after that everything returned to the old restrictive order. Nevertheless, this window in time, when a more liberal view prevailed, was of decisive importance for the development of anatomy, and consequently of medicine as a whole.

The greatest names of the Alexandrian school are Herophilus from Chaludon and Erasistratus from Ceos. Herophilus, often considered to be the pioneer of scientific anatomy, was the first to give a satisfactory anatomical description of the brain. He also realized that there are two different kinds of nerves, one that transmits motor impulses to the muscles and another that conveys sensory impulses from the periphery. Erasistratus concluded, based on his studies of the comparative anatomy of the brain, that the human intellectual superiority over animals was due to the development of the human cerebrum. Furthermore, he rejected Hippocrates' idea of the four body humors and instead maintained that the arteries did not contain blood but something that he called pneuma, which originated from the air in the lungs and was essential for the proper functioning of the organism. The fact that if one opens an artery in a living individual, a red, pulsating jet of blood gushes out — not air or pneuma — he explained in the following way. There must exist countless connections between arteries and the accompanying veins (Erasistratus called them anastomoses), which are normally closed. When an artery is injured, the anastomoses immediately open and the artery consequently becomes filled with blood from the adjacent vein. This idea of the arteries containing pneuma and not blood, which incidentally did not originate with Erasistratus but is considerably older, remained part of medical theory until Harvey demonstrated the circulation of blood in 1628.

Roman medicine: Compared to Hippocratic and Alexandrian medicine, Rome during the republican centuries was primitive indeed. The glorious exception was of course Roman hygiene with the admirable system of aqueducts carrying water to the eternal city and the extensive sewages, including the famous Cloaca Maxima. To these achievements should be added the organization of military medicine with hospitals for sick and wounded legionaries built in garrisons along the borders of the empire. In terms of medical theory and practice, however, the Romans were completely under the influence of religious and magical ideas about gods and demons as the cause of disease. At the same time, the number of Greek physicians in Rome continued to grow and even if most of them were imported slaves in the households of rich patricians, some were free entrepreneurs and very successful practitioners with patients in the upper classes.

The patrician Aulus Cornelius Celsus was the only native Roman, who made a great name for himself in medicine. His fame, however, rests on his systematic compilation of all the medical knowledge of antiquity rather than on any original contributions of his own. Celsus' medical writings, his *De Medicina* comprise eight books in a Latin that has aroused admiration and been considered a model over the centuries. The first book deals also with medical history, and much of what we know, for instance about Alexandrian medicine, we owe to Celsus. He is of course strongly influenced by Hippocratic ideas, and like Hippocrates he is at his best when he deals with surgery. For some reason, perhaps because they were in Latin and not in Greek like most medical literature, his writings were overlooked by contemporary physicians. It would take until the renaissance before *De Medicina* was rediscovered and became one of the first medical books to be printed. It was a widely used textbook and it may be significant that the great medical iconoclast Theophrastus von Hohenheim, more than a millennium after the death of Celsus, chose to be known under the name of Paracelsus.

If medical fame had been slow in coming to Celsus, the other great name in Roman medicine, Galen, had no such problem. Unlike Celsus he was a physician and practiced medicine with great success among the aristocracy of Rome. Like most doctors in The Roman Empire he was Greek, born in 130 AD in Pergamum, an important city of learning in Asia Minor. Galen started his medical education in Pergamum, then travelled extensively to different medical centers in the Eastern Mediterranean, including the most famous of them all, Alexandria. In 161 AD, he settled in Rome and started a very successful practice that included even the emperor-philosopher Marcus Aurelius himself. He was made court physician and given the task to supervise the health of Commodus, the heir to the throne. This gave him time both for research and extensive writing, including anatomy, pharmacology, physiology, pathology and all aspects of clinical medicine. He was essentially a follower of Hippocrates and devised an elaborate system where changes in the four body fluids, with their different properties of being cold, warm, moist or dry, could neutralize or reinforce each other according to a rather complicated scheme. He relied heavily on bloodletting, but also used a number of drugs which he considered to be cold, warm, moist or dry. For instance, pepper was warm and therefore counteracted an increase of the cold phlegm, while opium was cold and had the opposite effect of pepper.

What made Galen unique in Roman medicine and sets him apart from everyone else including Celsus, is his remarkable work as an experimentalist. He was active over a wide field of basic medical sciences, but his most important results were gained in anatomy and physiology. In the days of Galen, dissections of human bodies were no longer permitted and he had to be content with working on animals. In order to facilitate the extrapolation of his findings to humans, he concentrated on monkeys and made a number of important observations. Unfortunately, his attempts to generalize from monkeys to man resulted in mistakes and because of his almost Godlike position in the Middle Ages, these mistakes were not corrected until Vesalius challenged his authority during the renaissance.

The most impressive of all Galen's scientific achievements are his studies in neurophysiology. He showed that when the spinal cord of an experimental animal was transected at different levels, the resulting paralysis was more extensive the higher up the cord was cut. Furthermore, cutting half the cord and leaving the other half intact gave paralysis of only one side, corresponding to the level of injury. The prodigious writings of Galen represent the climax and summary of the whole of ancient medicine, and it is mainly through Galen that the teachings of Hippocrates have exerted their enormous influence for over two millennia. Undoubtedly this has to do with the tendency of the Middle Ages to view their own culture as inferior to that of ancient Greece and Rome, and the feeling that the renaissance represented the resurrection of the antique culture. However, Galen's own personality may also have played a role here. In everything that he writes, one feels his never-failing conviction of the infallibility of his own opinions. His complete lack of modesty is staggering and one can understand if to an admiring posterity his writings seemed almost as irrefutable as the unshakable dogma of the Catholic Church. When he died in 200 AD, he could certainly look back on a life's work that is unique in the history of medicine, both in its scope and in terms of its lasting influence.

Islamic Medicine

After the division of the Roman Empire in two in 395 AD, the destinies of the two empires were very different. While the Western Empire was soon engulfed by the maelstrom of barbarian invasions, the Eastern Empire or Byzantium, would last for more than a thousand years before it succumbed to the onslaught of the Ottoman conquerors in the 15th century. Periodically, it was the dominating power in the Eastern Mediterranean and during its long existence, Byzantium played an important role as the keeper of the Greek cultural legacy. This was true also in the medical field, and the writings of the Byzantine physicians maintained Hippocrates' tradition of the body fluids and their paramount importance, as well as the high standard of Hippocratic surgery. In this way, they became instrumental in saving the medical legacy of ancient Greece from being irretrievably lost in the cataclysm that ended the Roman Empire.

The astounding Arab conquest of a considerable part of the civilized world took less than a hundred years and had by the middle of the 8th century resulted in an empire that stretched from Spain and the Pyrenees to the Indian subcontinent and the Himalayas. The culture of the Arabian conquerors was often inferior to that of the subjugated peoples, but the Arabs were willing to learn. In fact, they showed themselves exceedingly capable at assimilating all that they found desirable in these foreign cultures, in particular natural sciences and medicine. They were aided in this endeavor by religious strife among the Christians in Byzantium. The patriarch Nestorius, who taught that the blessed Mary, as a human being, could not be the mother of God, had been deposed and expelled from Constantinople in 431 and his followers, the Nestorians, driven into exile. In this way many learned men found their way to Mesopotamia and Persia, countries that would later be overrun by the Arabs. The Nestorians translated Greek scientific and medical writings into Syriac, and these texts were then translated from that language into Arabic. Thus, the Arabs came in early contact with the work of Aristotle, Hippocrates and Galen. Several of these works have in fact come down to us as translations into Arabic and would otherwise have been lost.

Islamic medicine has its origin in Baghdad under the Abbasid caliphs, but later Cairo in Egypt under the Fatimids also became an important medical center, and in the 12th century several outstanding Arab physicians were active in the Western caliphate of Spain. It has often been said that the great historic importance of Islamic medicine is to have served as a refuge for the Hippocratic tradition, from where it could later recapture its rightful position in the medicine of Western

Europe. At the same time, historians have tended to depreciate Islamic medicine in terms of original contributions of its own and seen it only as the keeper and defender of a great tradition. In what follows, it will appear that this is not the whole truth.

Rhazes

Two names completely dominate medicine in the Eastern caliphates — Rhazes and Avicenna. The former was born in Persia around 865 in the town of Ravy and he died there sometime between 920 and 930. He worked as physician-in-chief at the great hospital that the caliphs had built in Baghdad which in every respect was far superior to the primitive hospitals of Western Europe at this time. Rhazes, who had wide intellectual interests, wrote more than 200 treatises on medicine and philosophy, but many of these have been lost. His most famous work, *A Treatise on Smallpox and Measles*, was translated into Latin and a number of other languages and enjoyed enormous popularity over the centuries. There had, in fact, been little room for the infectious diseases in Hippocratic medicine with its emphasis on the disturbance of the body fluids as the cause of all illnesses and Rhazes is the first author to give a clear description of smallpox and measles. At the same time it must be said that his treatise is very much colored by his Hippocratic thinking and it is doubtful if he recognized the infectious character of these diseases, which were probably very widespread in his day. Rather, he saw them as variations of the same basic illness and in a truly Hippocratic way considered them to be an adjustment of the infant's body fluids to life outside the womb. It is clear from his description of smallpox that this was in his opinion a fairly banal children's disease that practically nobody escaped. In later centuries, this almost idyllic view of smallpox would change drastically.

Avicenna

In many ways Avicenna (980–1037) can be considered as the Islamic counterpart to Galen in terms of the enormous influence that he came to exercise over innumerable generations of physicians, not only in the Islamic world but also in Europe. He was born near Bokhara in Persia and belonged to an aristocratic family of considerable wealth. His education included all the learning of the time, not only in medicine and the natural sciences but also in philosophy and jurisprudence. Avicenna wrote extensively on a number of subjects, but in what follows we restrict ourselves to his medical writings. His success as a compiler and commentator of Islamic medicine, which mainly meant the teachings of Hippocrates and Galen,

Uroscopy.
Courtesy of Bonnierförlagen, Stockholm, Sweden.

was unprecedented. The most renowned of his many books was his *Canon* that was received as a revelation even in the Christian countries in Europe.

Avicenna can be seen as the great advocate of ancient Greek medicine, but there is at least one field where he goes far beyond the teachings of the old masters. His *uroscopy* can indeed be said to represent Hippocratic medicine taken to its extreme. The visual inspection of the urine became by far the most important clinical method of examination in the Middle Ages. It was built on the old Hippocratic teachings about the four body fluids and the idea was that since the urine must obviously be derived from them, it should mirror their condition. It was with the triumphal progress of Islamic medicine over the world that uroscopy came to occupy its dominant position in medieval medicine. This can be seen also in numerous paintings of the doctor lost in deep thought while contemplating the inevitable urine-containing flask that he holds up against the light. To people in the Middle Ages, that flask came to be a symbol of the physician in the same way that the stethoscope is today.

It is understandable if this simple but highly ingenious method, that allowed of a detailed diagnosis without the examining doctor actually coming in physical contact with his patient — an important consideration during the plague epidemics of the Middle Ages — became immensely popular in the medical profession. Uroscopy spread like a religious revival all over the world, Avicenna was its prophet and his *Canon* its Holy Bible.

The Revival of Western Medicine

We know very little about medicine in Western Europe during the early centuries of the Middle Ages. However, there is every reason to believe that this was indeed a dark period in the history of medicine, when the knowledge of Hippocrates and Galen was lost in the political and social convulsions that followed on the fall of the Western Roman Empire. The first indication of a medical revival in Western Europe is the founding of a medical school in Salerno, probably some time in the 9th century. Salerno lies at the Tyrrhenian Sea near Naples, in a region rich in ancient monuments that remind us of the strong Greek cultural influence in this part of Italy. It is easy to understand that Greek medicine should come to play an important role in the Salerno school, whose faculty characteristically enough was known as Civitas Hippocratica. At the same time the influence of Islamic medicine was strong in this region, and nearby Sicily was under Arab rule until the Norman conquest in the second half of the 11th century.

Salerno became a center for the translation into Latin of Arabic treatises based on the writings of Hippocrates and Galen, thereby completing the long detour of Hippocratic medicine from Greece and Rome over the Arab Empire and back again to Europe. The leading name here was Constantinus Africanus (1010–1087), who owed his name to having been born in Carthage (according to another tradition he was from Sicily). Undoubtedly, the encounter with Islamic medical literature through the efforts of Constantinus Africanus and others was a revelation to Western physicians, and from now on the prestige of Arab medicine continued to grow in Europe.

By far the most popular and widely read of all medical writings from Salerno was a versified treatise on dietetics and personal hygiene called *Regimen Sanitatis Salernitatum.* It was first printed in 1480 and would pass through over 300 editions. From the original Latin, it was translated into a number of languages and it greatly influenced the medical thinking of ordinary people. It contributed, for instance, to the dissemination of the doctrine of the four temperaments, which became immensely influential.

The Middle Ages witnessed the founding of the first universities; perhaps the greatest achievement of that period from an intellectual point of view. It is difficult to decide which university can rightly claim to be the oldest, but Bologna has always been considered a strong candidate — Salerno had by and large remained exclusively a medical school. Outside Italy, the University of Montpellier could pride itself on its famous medical faculty and was remarkable also because of its open-mindedness in admitting both Jewish and Islamic students. The University of Paris was considerably less generous in its attitude, owing to the strong influence of the Catholic Church. Generally speaking the medieval universities emphasized the study of theology and this remained by far the most important faculty, followed by philosophy and jurisprudence, while medicine and the natural sciences held more modest positions.

Infectious Diseases

Perhaps the greatest original contribution of medieval medicine is the recognition of infectious diseases as an important part of the medical panorama. The Middle Ages had a vast and terrifying experience of devastating epidemics such as the plague in the 14th century, that is vividly described in the famous introductory chapter of Boccaccio's *Decameron.* It struck Florence in 1348 and in an unmistakable way demonstrated its contagious nature, spreading like a brush fire and wiping out the majority of the inhabitants. The authorities seem to have realized early on the true nature of the disease and drastic measures were taken to prevent it from spreading to uninfected areas. The Adriatic port of Ragusa, for instance, required all travellers to be isolated outside the city for a month before being admitted. Later this period was considered insufficient, and the time was extended to 40 days, "quarante giorno" in Italian, giving rise to the term quarantine.

By far the best known example of the isolation of patients to prevent a contagious disease from spreading are the brutal measures taken against the lepers. This disabling chronic disease was found all over Europe and had been known since antiquity. In the Middle Ages it was clearly perceived as infectious, and the consequences of being diagnosed as a leper were terrible also from a social point of view. The leper became an outcast shunned by everyone, had to live in special leper hospitals or colonies, and was required to carry a rattle that warned of his approach.

In addition to epidemic diseases like plague, measles and smallpox, there came at the beginning of the renaissance a violent outbreak of what seemed to be a new infectious disease, syphilis. This disease is generally believed to have been brought to Europe by sailors and soldiers returning from the newly discovered Americas. Alternatively, it might have been endemic in the old world and the explosive outbreak could have been caused by the drastic increase of promiscuity at the end of the Middle Ages and the beginning of the renaissance, when the Church lost much of its hold on people.

The leading Italian poet and master of Latin verse, Girolamo Fracastoro (1478–1553), was also a physician and one of the great pioneers of what would 300 years later be known as microbiology. In his epic poem about the shepherd Siphilo, who had offended Apollo and was punished by the revengeful God with the terrible disease, Fracastoro described the symptoms of syphilis (the disease was named after the hero of his poem) in considerable detail that testifies to his familiarity with it. However, his interest in infectious diseases was not limited to syphilis. He noted, for instance, that tuberculosis could be transmitted by bedclothes and garments previously used by the sick. His studies led him to believe that contagious diseases were caused by invisible particles, which he called "seminaria contagiosa" and considered to be self-replicating and specific for each disease. Fracastoro's theories were far ahead of his time and never had any real influence on contemporary medicine. Nevertheless, this poet of the Italian renaissance is one of the great sages of medicine, remarkable because his theories were not pure speculations, like so many others of this time, but built on thorough, critically evaluated clinical observations.

Lepers greeting death as their deliverer. Fresco in Camposanto, Pisa.
Courtesy of Bonnierförlagen, Stockholm, Sweden.

Mondino de Luzzi, called Mundinus (1270?–1326).
Courtesy of Bonnierförlagen, Stockholm, Sweden.

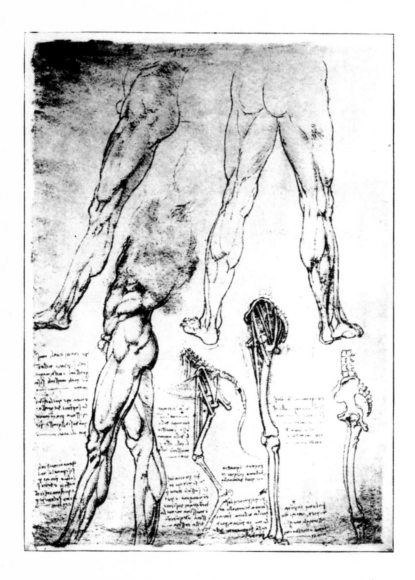

Anatomical drawings by Leonardo da Vinci (1452–1519).
Courtesy of Bonnierförlagen, Stockholm, Sweden.

A New Anatomy

In the 14th century, the very strict attitude of the Church towards the dissection of human bodies would seem to have become somewhat more liberal, and there was a revival of the study of anatomy. The leading anatomist at the University of Bologna was Mondino de Luzzi, called Mundinus (1270?–1326), who performed a number of dissections. The truth of the matter is, however, that it would be more correct to say that dissections were performed in his presence by assistants without any theoretical background, while the great professor held aloof from the odorous corpse and lectured from the books of Galen. As a consequence, when his great *Anathomia* was completed in 1316, it repeated all the mistakes found in the writings of Galen that depended on false analogies between human anatomy and the findings in animals.

Thus, the leading anatomists of the Middle Ages had not been able to add much to the inheritance from the divine Galen, but with the dawn of the renaissance this would change radically. It is easy to understand that the great artists of this period, when the human body was depicted more freely and lovingly than it had ever been since the golden age of Greek civilization, should be fascinated by anatomy. This is certainly true of men like Michelangelo and Dürer, but the incomparable pioneer was Leonardo da Vinci (1452–1519). Unfortunately, Leonardo's great anatomical contributions were never revealed to contemporary medicine. His secretiveness prevented him from publishing his drawings, and it was instead a young physician from Brussels, Andreas Vesalius (1514–1564), who revolutionized anatomy.

Vesalius was educated at the medical faculty of Louvain and later in Paris, and he has given a scornful and quite amusing description of the teaching of anatomy there. Having been appointed professor of anatomy in Padua already as a young man, he personally led the dissections and demonstrations and attracted great audiences to his lectures. In the beginning of his career he was very much an admiring and unquestioning follower of the infallible idol from Pergamum. When his findings in human corpses differed from Galen's writings, he simply dismissed his own observations as aberrations from the normal. However, when he later came to dissect a monkey and found a detail in its skeleton that exactly agreed with Galen's description of human anatomy in this respect, the truth dawned on him. Galen, who was not permitted to dissect human bodies, had simply extrapolated from his findings in animals, mostly monkeys, and some times he had guessed wrong — human anatomy did not always agree with that of monkeys. This put an end to his blind faith in Galen and provided the impulse for his life's work

Andreas Vesalius (1514–1564).
Courtesy of Bonnierförlagen, Stockholm, Sweden.

and one of the most important books ever published in medicine — *De Humanis Corporis Fabrica.*

The amount of work that must have gone into the making of this book with its 300 illustrations boggles the mind. We do not know who made the pictures, but they certainly were of a quality that had never been seen in the medical world before. The illustrations are extremely well integrated with the text in a way that testifies to the intimate collaboration between Vesalius and whoever made them. In the year 1542, Vesalius sent the manuscript and the hundreds of heavy blocks containing the woodcuts on the dangerous roads over the Alps to the printer Johannes Oporinus in Basel. Next year came the first edition of this truely remarkable book, important not least because it for the first time freed a medical subject, anatomy, from the stifling authority of Galen.

Vesalius' attack on the time-honored teachings of the classic authors was a bitter medicine for many of his colleagues to swallow and they fought back mercilessly. This backlash of what he must have perceived as mindless traditionalism and spiteful hostility hurt Vesalius deeply. He even burnt some of his manuscripts and left his chair in anatomy at Padua to become court physician to the Spanish King Philip II. The stiff formalism and bigotry of Philip's court can hardly have been congenial to Vesalius. In 1564 he went on a pilgrimage to the Holy Land, possibly to reconcile himself with the Catholic Church whose wrath he may have incurred by his acid wit. On his journey the ship he travelled in was wrecked at the little island of Zákinthos where he took ill and died.

There were a number of outstanding anatomists in Italy during the renaissance, men like Falloppio, Fabricius ab Aquapendente, and Bartollomeo Eustachi, but none of them can compare with Vesalius in importance. The reason is that as the first he dared to challenge the never questioned authority of Galen and thus he freed anatomy from the yoke that had stifled it. In fact, his break with the teachings of the old authorities would have far-reaching consequences also in other fields of medicine.

Harvey and the Blood Circulation

William Harvey's discovery of the blood circulation and the function of the heart shook the foundations of a medical doctrine that went back all the way to Hippocrates and Aristotle. It was believed that blood was formed in the liver and then transported in the Vena Cava to the heart. The main function of this organ was to act as the source of body heat. Furthermore, the activity of the

William Harvey (1578–1657).
Courtesy of Bonnierförlagen, Stockholm, Sweden.

right ventricle gave a kind of to-and-fro undulating movement to the blood in the venous system in order to keep the body fluids well mixed. The arteries, on the other hand, were filled with a special kind of air (pneuma), and what little blood they contained came from the right ventricle through the ventricular septum that was believed to be to some extent permeable to blood. The pneuma, or "animal spirit" as it was also called, was sucked in from the lungs and distributed through the arteries to different parts of the organism, where it was necessary for the vital processes. Because of the consumption in the tissues of the arterial pneuma as well as nourishment from the blood, there was a slow movement of the content in both arteries and veins from the heart to the periphery. Thus, it is fair to say that when Harvey took up his medical studies in Padua at the end of the 16th century, medical science had got everything regarding the heart and its function hopelessly wrong. This was in spite of the fact that by then the lesser circulation, i.e. the movement of the blood from the heart to the lungs and back again, had already been discovered.

It is difficult to imagine a more unlikely medical pioneer than the Spanish physician and religious mystic Michael Servet (1511–1553). He had been in exile, a fugitive from the inquisition, ever since at the age of 20 he wrote a heretic pamphlet where he denied the Holy Trinity. Servet had settled in France, where he had a successful medical practice near Lyon, when in 1553 he published his *Christianismi Restitutio* (the restoration of Christianity). Here his discovery of the lesser circulation is mixed in the most amazing way with horrible blasphemies concerning the Trinity, the sacrament of baptism and a row of other dogmas. He became an outlaw in all Christian countries, Protestant as well as Catholic, and was finally hunted down and burned at the stake in Geneva together with his ungodly book. Such was the hatred that Servet had aroused in all Christendom that even his discovery of the lesser circulation and the change of the color of the blood as it passed through the lungs, was suppressed. Instead, it was the Italian anatomist Realdo Colombo who, six years after Servet's death, claimed to have discovered the path of the blood from the right ventricle to the lungs and from there to the left auricle and ventricle of the heart.

William Harvey (1578–1657) came from a rather wealthy English family and could afford to study at the most famous of all medical schools, that in Padua. Here he must have heard Fabricius ab Aquapendente lecture about his great discovery, the valves that he had found in the veins. Fabricius completely misunderstood their function (like everyone else he thought that the blood of the veins flowed from the heart towards the periphery), but maybe his young English student already at that time had realized the true explanation. The movement of the blood in the veins is from the periphery towards the heart and the valves are there to prevent it from flowing in the opposite direction.

Marcello Malpighi (1628–1694).
Courtesy of Bonnierförlagen, Stockholm, Sweden.

In any case, when he returned to England and became a lecturer at the Royal College of Physicians, Harvey's lecture notes reveal that at least in 1616 he had worked out the main features of the blood circulation. However, it was only after having performed innumerable experiments on all kinds of animals that in 1628 he published his short book *Excercitatio Anatomica de Motu Cordis et Sanguinis in Animalibus* ("An Anatomical Investigation of the Movements of the Heart and the Blood in Animals").

In this book, Harvey presents a number of arguments to sustain his heretic ideas, for instance the fact that the arteries contain blood and not pneuma. Characteristically, an important point is that the divine Galen himself had considered the arteries to be filled with blood. Harvey argues that the heart in a very short time pumps out a volume corresponding to all the blood present in the body. Obviously, the only reasonable explanation is that the blood has a circular movement that continuously brings it back to the heart again. He then goes on to give a correct description of the blood circulation, including the fact that the construction of the valves between the right ventricle and the pulmonary artery, and the left ventricle and the aorta, respectively, is such that the blood can only flow in one direction, from the heart and into these vessels, consistent with the suggested circulation.

Harvey had thought that his heretic ideas would cause a lot of hostile criticism, but on the whole they were fairly well received. The great French philosopher René Descartes, with his view of the living organism as a very complicated machine, was supportive and considered that Harvey's blood circulation fitted quite nicely into Descartes' own thinking. When the Italian anatomist Marcello Malpighi in 1661 could demonstrate a system of fine tubes (capillaries) through which the blood could be seen to flow from the arteries to the veins, Harvey's model of the blood circulation was complete. It is not unreasonable to claim that this discovery is the most revolutionary in the history of medicine.

ALCHEMY AND THE DAWN OF CHEMISTRY

The Gold-Makers

Alchemy does not have much standing as a science these days. Rather it is seen as pure superstition and its practitioner, the alchemist, as an ignoramus, in many cases even an imposter, whose only ambition was to swindle his poor victims by taking advantage of their credulity and greed. Nevertheless, alchemy played a significant role in the leading human cultures since time immemorial and it can with some justification be considered as the origin of modern chemistry.

A central idea of alchemy was the notion that all metals can be transformed into each other by a series of more or less complicated chemical operations. The rationale for this belief was the generally accepted concept of a prime matter suggested by Aristotle. Metals were very closely related to each other since they had all arisen as mixtures of two vapors, one dry or smoky and one moist, that originated from the earth. These vapors were later identified by the alchemists as sulfur (the dry vapor) and mercury (the moist vapor). Thus, all metals were mixtures in different proportions of sulfur, which signified combustibility, and mercury, which symbolized fluidity and changeability. The idea that they could be changed into each other (transmutation) by suitable procedures was therefore close at hand. It is here that we encounter the concept of the philosopher's stone, sometimes also envisaged as an elixir or potion, which acted as a ferment in the process, in the same way as leaven did in the making of bread or the brewing of ale.

Studying the history of alchemy is not always an edifying experience. The old alchemists often appear as a bunch of unprincipled liars with their tall stories of magic transmutations where a grain of the philosopher's stone transforms large quantities of base metals such as lead or mercury into pure gold. At the same time, they can seem almost unbelievably credulous and naive when they squander vast amounts of money in the pursuit of the elusive stone that will hopefully make their fortune. Even if they should eventually turn out to be successful, their troubles are far from over. Impecunious princes with empty coffers and an insatiable lust for gold lie in ambush for them and think nothing of putting a poor alchemist on the rack to squeeze his secrets out of him. Another scenario might be that an irascible prince, who considers himself to have been swindled by the gold-maker, terminates their relationship by hanging the former protégé from the nearest tree.

How is it then, that we nevertheless can see alchemy as the dawn of modern chemistry? It has to do with the fact that the alchemists were the first to set up real chemical laboratories in which they worked out many of the basic methods that are still used in modern chemistry. Furthermore, alchemy represented all the knowledge of inorganic chemistry at the time and must have played a significant role in the development of metallurgic processes used in the economically

Alchemists trying to obtain the philosopher's stone.
Source: Author's collection.

Distillation is a method we probably owe to the alchemists.
Source: Author's collection.

important mining industry. The alchemist's laboratory was a place that appealed to people's fantasy, and numerous woodcuts and prints that depict it have come down to us. The central piece of any such laboratory was always the furnace that provided the heat required for most of the processes in alchemy, whether it was the laborious and often incredibly time-consuming procedures for obtaining the indispensable philosopher's stone or the transmutation itself, where the stone acted as the necessary ferment. One such operation, where the furnace was a prerequisite, was distillation, a method that we probably owe to the alchemists. In any case, it was perfected during the heyday of alchemy in the Middle Ages. Most pictures of the alchemist's laboratory show an apparatus for distillation, often of a rather sophisticated design. We know for instance that some of these could produce better than 90% pure alcohol by the distillation of wine.

It is striking how the history of alchemy runs parallel to that of medicine. Its practitioners were often physicians and when classical medicine took refuge in the Islamic world from the upheavals and cultural decline in Western Europe, which followed on the fall of the Roman Empire, it was accompanied by alchemy. For many centuries, both medicine and alchemy survived and flourished in the Islamic countries and several of the leading names in Islamic medicine were also alchemists, the best known being Rhazes and Avicenna. It is worth noting that these medical paragons seem to have been more interested in the supposed ability of the philosopher's stone to cure disease, than in its use to transmute base metals into gold. When medicine eventually returned to Western Europe from the Arab Empire, alchemy followed in its footsteps.

The leading Islamic alchemist was Jabir ibn Hayyan, who was active in the 8th century AD. However, his numerous books do not necessarily all have Jabir as their author. Many of them may in fact have been written by others but attributed to Jabir because his great fame and authority as an alchemist reflected credit on them. A number of Jabir's books, the most important one being his *Summa Perfectionis*, were translated into Latin. At the same time, his name was latinized to Geber and as "Geber, the most famous Arabian Prince and Philosopher", he came to dominate the alchemy that now began to flourish in medieval Europe. His writings, or rather the books attributed to him, contain the essence of Arabic alchemy including the fanciful theories about the nature of metals and their transmutation into each other and the role of the philosopher's stone in the process. But this was not all, they also described a variety of chemical methods, for instance the preparation of strong inorganic acids such as sulfuric acid and nitric acid. Other important experimental innovations were the use of distillation for the separation of liquids and sublimation to convert a solid directly to a vapor. Regardless of how much credit should be given to Geber himself, there can be no doubt that what passes as his writings were of the greatest importance for making Islamic chemistry known in medieval Europe.

The 12th century represents an intellectual revival of Western Europe, and it is also the period when the practice of alchemy becomes more widespread. At this time, the European alchemist was typically a man of the church; it was after all among priests and monks that literacy was the rule rather than the exception, as was the case in the rest of the population. Such leading figures of the church as Albertus Magnus and Thomas Aquinas have been viewed as believers in alchemy, but it is doubtful if they actually subscribed to such claims as the transmutation of base metals into gold. Likewise Roger Bacon, with his profound interest in experimental science, has had alchemical treatises attributed to him. It is, however, an open question if they actually emanated from him. Nevertheless, there is no doubt that Islamic alchemy, when it took root in Europe, at the same time greatly promoted the study of chemistry there. Eventually, chemical concepts would also influence medical thinking and upset the seemingly unshakeable foundations of Hippocratic medicine.

Iatrochemistry — A Chemical Explanation

The history of medicine is full of controversial figures but Paracelsus, the founder of iatrochemistry, must surely be assigned a leading place in this gathering of eccentric geniuses. With the dawn of the renaissance, the time-honored wisdom of the classic authors was questioned in basic medical sciences like anatomy and later also physiology, but in clinical medicine Hippocrates and Galen ruled unchallenged, except for what seemed to be the ravings of a madman; at least that was the verdict of his shocked and offended colleagues at the leading medical schools in Europe.

Philippus Aureolus Theophrastus Bombastus von Hohenheim, who called himself Paracelsus to emphasize that he was the equal of the great classic authorities Hippocrates, Celsus and Galen — perhaps even their superior — was born in Switzerland in 1493 as the son of a Swabian physician. On his father's side, he came of a noble family but his mother was a peasant woman, and there is nothing aristocratic about Paracelsus, either in appearance or character. His father had studied medicine at the University of Tübingen and he was also a fair scholar in the humanities, so that the years that Paracelsus spent as an apprentice with him must have been of decisive importance both for the boy's intellectual and professional development. Already as a youngster, Paracelsus embarked on the

Paracelsus (1493–1541).

Courtesy of Bonnierförlagen, Stockholm, Sweden.

endless travelling around Europe that was to continue almost without interruption until his death. In his copious writings, he claims to have sought out people from all walks of life in order to learn about Nature and the treatment of diseases. At the same time, it is difficult to follow his studies at the many German universities he seems to have visited; for instance Tübingen, Heidelberg, Frankfurt, Cologne and Munich, to mention just a few. Nor were his studies limited to Germany. After two years in Ferrara, he was created a doctor of medicine there in 1515, or so he claims at least, but there are no records of this at the university. In any case, his experience of universities and their professors had left him with a deep hatred and mistrust of academic medicine, with its mindless traditionalism and uncritical accepting of the teachings of the classic authors. Just listen to him as he stands alone and defiant on the scene of Europe:

"My will shall prevail, not yours, gentlemen of Paris, of Montpellier and of Swabia. In the farthermost nooks the dogs shall one day piss on you! — My shoelaces are more learned than your Galen and your Avicenna. I will be King and the Kingdom will be mine!"

These opinions, that he expressed freely, cannot have endeared him to his colleagues in the medical profession and with one exception he never held a position at any of the many universities he visited. In 1526, he was appointed professor of medicine at Basel where he shocked the other professors by lecturing in German and not in Latin as custom required. Even worse, in the presence of his students he threw the revered books of the classic medical authors into the fire to demonstrate his complete break with the teachings that had dominated medicine for almost 2000 years. Characteristically enough, he soon found himself in acrimonious conflict with one of the councillors of the city over fees for medical services, and he resumed his incessant travelling.

He had completely rejected classic medicine and denounced its great figures in the past, with the exception of Hippocrates whom he saw as having been misrepresented by Galen and Avicenna. But what were his own views? What had he to offer instead of the all-explaining theory of the four body fluids? Paracelsus took the view that sickness and health could be explained, not by the relative proportions of the body fluids, but in terms of three chemical principles: *sulfur*, that signified combustibility; *mercury*, that symbolized fluidity and changeability; and *salt* that stood for solidity and stability. The delicate balance between these chemical principles in the body was essential for its normal functioning, and disturbances of the balance caused disease. This attempt to explain the normal functions of the organism, as well as its different diseases, in chemical terms would later be known as iatrochemistry (*iatro* meaning physician).

Undoubtedly, this theory was influenced by the interest of Paracelsus in alchemy, even if, like Rhazes and Avicenna, he was primarily fascinated by the ability of the philosopher's stone to cure disease, rather than its role in the making of gold. As we have seen previously, alchemy embodied all the knowledge of inorganic chemistry of the time and it is logical that Paracelsus was the first to introduce preparations containing heavy metals in the pharmacopoeia. This addition to the therapeutic arsenal became very popular and must have caused untold suffering over the centuries.

Even if the medical thinking of Paracelsus was to a large extent dominated by his chemical theories, it also included magical, not to say superstitious elements that illustrate to what extent he was still a child of the Middle Ages. He introduced the enigmatic concept of *Archeus* as a symbol of the principles that ruled the normal functions of the body and prevented it from succumbing to disease. When *Archeus* did not govern anymore, the body became ill, and it was the task of the physician to try to re-establish the benevolent rule of *Archeus*. A central principle of Paracelsus' is the idea that for every disease Nature has designed a specific remedy, and here he adopts the medieval doctrine of *signatures*. For instance, plants that in one way or another are reminiscent of a characteristic feature of the disease or its imagined cause, contain a principle that counteracts the disease. Thus, the yellow color of saffron showed that it could be used as a medicine against jaundice. These specific remedies, indicated as it were by the Creator himself, Paracelsus called *Arcane* and they played a central role in his therapeutic thinking.

Paracelsus had none of the social graces or the well-rounded personality that one tends to associate with the successful clinician. Nevertheless, he may have been a good doctor, appreciated by his patients. What we know for a fact is that his colleagues considered him a dangerous lunatic whose anarchistic views in medicine were anathema to the profession. However, if one wants to take a charitable view of his fantastic theories, they can be seen as the first attempt to explain physiological phenomena in chemical terms. In fact, one might perhaps consider the appearance of Paracelsus and his iatrochemistry as the dawn of physiological chemistry. This is consistent with the fact that he always emphasized that science is a necessary part of the medical education, including of course chemistry, which at this time also meant alchemy. Even so, he was very much a product of the transition period between the Middle Ages and the renaissance, and his teachings are an enigmatic mixture of superstition and a prophetic vision of a new medicine based on the natural sciences. When his long wanderings finally came to en end in 1541 at the inn *The White Horse* in Salzburg, he left an important legacy to posterity that would influence medicine for generations.

If Paracelsus can be seen as the pioneer of iatrochemistry, the Flemish aristocrat Johann Baptista van Helmont is its most famous protagonist. At the same time they were as different as could be. Paracelsus was uncouth and aggressive, and thought nothing of heaping the coarsest invectives over his adversaries at the leading medical faculties, while the aristocratic van Helmont was quietly sarcastic in his merciless criticism of Hippocratic medicine and its obsession with its bogeyman, the dangerous phlegm.

Johann van Helmont was born at Brussels in 1577, and like one of his main adversaries, the great Islamic physician Avicenna 600 years earlier, he was the scion of a noble family of considerable wealth. He showed wide interests both in science and philosophy during his studies at the University of Luvain, but he eventually settled for medicine, an unusual choice for a man of his birth, and received his doctor's degree in 1599. Unlike Paracelsus he did not practice medicine extensively but was mainly interested in its theoretical and philosophical aspects. He was widely read in the works of the great classical authorities, but he found very little there that he could agree with. In his collected writings, *Ortus Medicinae*, he aimed at creating one all-embracing system that would include the answers to all the problems of medicine. However, like so many other great spirits he is much more persuasive in his criticism than in his attempts to build new, grandiose theories of his own.

In Hippocratic medicine, phlegm emanating from the brain was thought to find its way into various organs in the body by many different routes, thereby causing a great number of illnesses, for instance of the lungs. The voluminous expectorates of what was obviously phlegm, that accompanied these conditions, was seen as irrefutable proof of the theory. Against this, van Helmont argued that it was impossible for such large volumes of phlegm to have passed from the brain to the lungs by way of the larynx, when a few drops of liquid accidentally swallowed down this way was enough to cause violent coughing. Instead, he maintained that the phlegm resulted from a local pathological process in the affected lung; it was the result of the illness, not its cause. This may seem self-evident to us, but in reality it was something quite revolutionary. What van Helmont did here was to stand classic medicine on its head, at the same time deeply offending his colleagues among the physicians.

Perhaps even more shocking was his complete rejection of the mainstay of all therapy, bloodletting. According to van Helmont, it drained the patient not only of his blood but also of his strength, his hopes and the money in his purse. As an example, he cites the recent death in ague of the Spanish King's brother and charges that the physicians in attendance had not left a drop of blood in their poor victim, thereby obviously causing his premature death. How many murders, exclaims van Helmont, committed by physicians have not gone unpunished because the perpetrators could point to the doctrines of classic medicine to vindicate themselves. One can understand that he must have been a constant vexation to his medical colleagues.

Johann Baptista van Helmont (1577–1644).
Courtesy of Bonnierförlagen, Stockholm, Sweden.

Franz de le Boë, called Sylvius (1614–1672).
Courtesy of Bonnierförlagen, Stockholm, Sweden.

It is easy enough to agree with van Helmont when he rejects the four body fluids as the all-explaining theoretical foundation of medicine and condemns bloodletting as the destroying angel of conventional therapy, but the theories that he himself proposes in his *Ortus Medicinae* are equally fantastic to the modern reader. He revives Paracelsus' old concept of *Archeus* as the ruler of the human organism. Strangely enough, *Archeus* is believed to have his seat in the stomach from where he directs all the functions of the body. Exactly like the kings and princes that van Helmont could observe in his own age, *Archeus* might react with uncontrollable rage when provoked. He would then avenge himself by allowing the acid content of the stomach to attack different organs in the body, thus causing disease. Johann van Helmont's *Archeus* is really very far from the benevolent ruler that Paracelsus had envisaged, and maybe this theory was primarily the result of van Helmont's own chronic acid indigestion.

In spite of all these unfounded speculations, we must not forget that van Helmont also made important chemical discoveries. For instance, he described the formation of what he called *gas sylvestre* (carbon dioxide) as the result of the burning of charcoal and the fermentation of must. He also coined the term *gas* as distinct from air and water vapor. Johann van Helmont was fascinated by ferments and fermentation and he believed that all processes in the organism were caused by ferments that converted food into living flesh in six steps.

Johann van Helmont was considered a madman by the medical profession in the same way that Paracelsus had once been. Characteristically enough, after his death in 1644 van Helmont's colleagues tried to explain both his decease and his alleged madness by the fact that he had always rejected bloodletting, that sure guarantee of a long and healthy life. Together with Paracelsus he is the leading figure of iatrochemistry, but it would take an entirely different kind of personality to popularize the new ideas, than these two deeply original figures.

Franz de le Böe, called Sylvius (1614–1672) had all the personal qualities required of a successful clinician and teacher, that both Paracelsus and van Helmont so obviously lacked. He was born in Hanau in Hesse, got his medical degree in Basel and was appointed professor of medicine at the University of Leyden in 1658. Here he reformed the medical education and introduced systematic clinical training of the students by demonstrations of patients in the wards, following teaching principles first developed in Italy.

In his main work *Praxeos Medicae Idea Nova*, where the very title emphasizes Sylvius' claim to be a pioneer, he avails himself of several central ideas emanating from Paracelsus and van Helmont. For instance, he maintains that all processes of life depend on chemical reactions that he calls fermentations and likens to the well-known fermentation of sugar to alcohol. In his opinion, the breakdown of food, brought about mainly by ferments present in the saliva and the pancreatic juice, eventually produces material that can be taken up by the blood while the remainder is excreted as feces. Provided that we take the term "ferment" to signify what we today call enzymes (not necessarily a valid assumption as will be discussed below), these ideas would certainly appear to come close to our present concepts.

Sylvius attempts to explain sickness and health as depending on the delicate balance between acid and alkaline end-products of what we would call metabolism. Ideally, they are in a kind of equilibrium with each other and the organism is healthy. However, if this balance is disturbed, an alkaline or acid excess (Sylvius called this *acrimony*) may result and cause disease. All illnesses can therefore be classified as either alkaline or acid in nature. To give an example, syphilis that was a very common disease at this time, was caused by an excess of acid and phlegm. Iatrochemist as he was, Sylvius was nevertheless not above invoking Hippocrates and Galen, making use of their body fluids when necessary.

Ever since the days of Paracelsus, mercury was used to treat syphilis and one of Sylvius' followers, Stephan Blankaart, gives a fanciful and in many ways revealing explanation of how this treatment works. He suggests that the acid principle, which causes syphilis, can be envisioned as consisting of sharp-edged particles of phlegm that are taken up on the surface of little balls of mercury during the treatment. Every time these particle-coated mercury balls pass through the salivary glands they are stripped of their coat and the acid particles are eliminated with the saliva. It should be pointed out that the treatment with mercury causes a copious salivation.

Blankaart's embroidering on Sylvius' theories is of interest because it illustrates how far removed the ideas of iatrochemistry actually are from the reality to which they seemingly come so close. It is tempting to over-interpret the similarity in terminology, for instance to equalize "ferments" with the modern concept of enzymes, and ascribe an insight to the 17th century iatrochemists that they did not possess.

The Man on the Balance

The pioneer of iatrophysics (or iatromechanics, as it was also called) was the Italian physiologist Santorio Santorio (1561–1636), who introduced quantification as an important method in studying physiological processes. He actually spent a

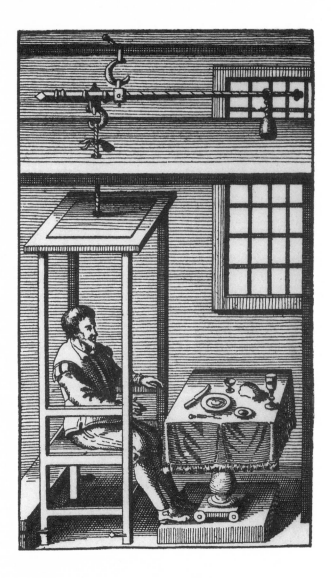

Santorio Santorio (1561–1636).
Courtesy of Bonnierförlagen, Stockholm, Sweden.

considerable part of his adult life on a balance of his own construction in order to quantify the variation in body weight and its relation to the amount of food and drink consumed and feces and urine excreted. Santorio consistently found that the weight of his food and drink considerably exceeded the total weight of his feces and urine. He correctly inferred that the body gave off some invisible substance (water vapor) and he called this phenomenon *perspiratio insensibilis* (invisible perspiration). After years of careful studies, Santorio became convinced that *perspiratio insensibilis* was of vital importance for health and disease. In fact, anything that reduced the amount of invisible perspiration caused disease and, conversely, everything that stimulated it was healthy. As prophylaxis, he recommended moderate exercise, in particular sedate dancing like in the popular saraband, a solemn Spanish court dance. To cure already manifest illnesses, Santorio prescribed somewhat more drastic measures, for instance treatment with sudorific preparations of sarsaparilla root and guaiacum that became widely used as a remedy for a number of diseases.

Generally speaking, iatrophysics/iatromechanics, that had many famous representatives at this time in the medical community, makes a somewhat naive impression on the modern reader with its mechanistic explanations of all physiological functions. Digestion was seen as a grinding down of the food to small particles; menstruation was a consequence of the upright position, which led to an accumulation of blood in the uterus that regularly relieved itself of the excess blood. The nerve impulse was explained as a movement of fluid in the hollow tubes believed to make up the nerves; fever was caused by the increased velocity of the bloodstream in the vessels causing friction against their walls with the production of heat *et cetera*.

At the same time, the principles of iatromechanics were in harmony with the teachings of the leading philosopher of the period, René Descartes, who was of the opinion that the body could be seen as an incredibly complicated and sophisticated machine. It is easy to understand that these straightforward explanations by mechanistic analogies, in contrast to the somewhat obscure reasoning of the iatrochemists, should have much to recommend them to the physicians. As one might expect, iatromechanics became the leading medical theory during the 17th and 18th centuries.

MEDICINE AND CHEMISTRY IN THE SCIENTIFIC REVOLUTION

Speculative Medicine — The System Builders

The great progress of science and mathematics during the 17th and 18th centuries had a profound impact on the prevailing conception of the world. Names such as Robert Boyle, Isaac Newton, Gottfried Leibnitz, Carolus Linnaeus and Antoine Lavoisier bear witness to what a golden age this was. It is a paradox that this heyday of science coincides with a period in the history of medicine when a number of highly speculative physicians indulged in the building of veritable castles in the air, theories that had absolutely no real foundation in clinical observations or experimental results and that have vanished without a trace. Nevertheless, at the time these airy theories were influential to an extent that we find difficult to comprehend.

The All-Pervading Ether

Newton had suggested that there existed a medium that he called ether, which pervaded everything in the universe and consequently was part of all living organisms. The German physician Friedrich Hoffmann (1660–1742), an amiable and attractive man, who was a close friend of both Boyle and Leibnitz, seized on Newton's idea of the ether and with it as a foundation erected an edifice of grandiose theories that captivated many of his colleagues. Hoffmann thought that the ether, which in the human body he believed to be located mainly in the cavities of the brain (the brain ventricles) could find its way through the peripheral nerves and reach the "fibers," that he considered to be the smallest building blocks of the organism (corresponding in a way to the cells of modern biology). This flow of ether from the brain determined the degree of tension in the fibers. If there was a rapid flow of ether through the nerves to the periphery, the tension in the fibers increased and when the flow became too intense, the result was a spasm. On the other hand, when the flow was too slow a paralysis set in. All illnesses were seen as variations on this basic theme. Spasm in the fibers resulted in diseases of an acute nature, while paralysis of the fibers gave chronic conditions.

Another central idea of Hoffmann's was that certain organs are linked to one another, he called it sympathy, and that through this sympathy they could influence each other. For instance, there was sympathy between the stomach and the brain. The stomach often became the seat of inflammations, and because of the existing sympathy between the organs this inflammation, could spread to the brain. One is undeniably reminded of van Helmont's irascible Archeus that ruled the human body from his headquarters in the stomach.

Friedrich Hoffmann (1660–1742).
Courtesy of Bonnierförlagen, Stockholm, Sweden.

Georg Ernst Stahl (1660–1734).
Courtesy of Bonnierförlagen, Stockholm, Sweden.

As a consequence of his dualistic view of diseases being either spastic or paralytic in nature, Hoffmann managed with a very limited number of drugs, a blessing for his patients no doubt, compared to the involved prescriptions of his colleagues. His medicines were either what he called "tonica," which supposedly increased the tension in the fibers, or spasmolytic preparations to counteract the effects of excessive tension. To this should of course be added the indispensable bloodletting.

A Troublesome Recluse

Hoffmann occupied the chair of medicine at the University of Halle, and he had a famous colleague and mortal enemy among the professors there, Georg Ernst Stahl (1660–1734), who had worked out a medical system quite as fanciful as that of Hoffmann himself. Otherwise, Stahl was in every respect the opposite of his colleague and enemy. While Hoffmann was easy-going and lovable with lots of friends, Stahl was an intolerant and difficult recluse who very easily got into conflicts with people. He was the major proponent of a bizarre chemical theory about combustion involving the mysterious idea of something called "phlogiston." The phlogiston theory was widely accepted even by outstanding chemists and we will consider it in more detail later.

Stahl's upbringing had been strict and very religious and, consequently, the soul or "anima" dominated all his medical thinking. In his magnum opus, that characteristically enough has the title *Theoria Medica Vera*, he introduced the idea that anima rules all the vital processes of the body. Anything that interfered with the benevolent rule of anima could result in a disturbance of the normal body functions, the most important aberration being an excess of blood, "plethora," that was the main cause of disease. Consequently, bloodletting and purging played a major role in Stahl's therapy. He considered fever to be an attempt by anima to cleanse the organism of whatever caused the disease, and he was therefore fanatically opposed to the use of quinine since it reduced the fever and thus obstructed the efforts of anima to cure the malady.

Magnus Boerhaavius

Hoffmann and Stahl had been successful enough in their efforts to propose all-embracing medical theories, but their Dutch colleague, Hermann Boerhaave (1668–1738), who was active in Leyden, far surpassed them. He was the leading clinician of his time and his many admiring followers always referred to him as Magnus Boerhaavius (the great Boerhaave). He must have had a very attractive personality that captivated patients and pupils alike, but the theories he advanced in his *Aphorisms* vied with those of Hoffmann and Stahl in terms of fanciful ideas. For us, who have not had the privilege to know the great clinician in Leyden personally and experience the remarkable charisma that he must have possessed, it is perhaps difficult to comprehend the enormous influence he had on clinical medicine in his day. However, it is notable that he had a number of outstanding pupils who loved and revered him.

Boerhaave had been strongly influenced by the discoveries of Harvey and Malpighi. He was particularly fascinated by the capillaries that connected the arteries with the veins and they became the mainstay of his pathology. The idea was that it required a major effort for the blood to pass through the tiny capillaries and anything that made the passage more difficult, for instance a thickening of the blood, could cause disease. As an example, he pointed to the classical symptoms of inflammation — reddening, increased temperature and swelling — that could clearly be explained by the blood having difficulties getting through the capillaries. The pulsations that the patient felt in the inflamed tissue meant that the blood was trying to force its way through the clogged vessels.

He was somewhat vague when he tried to explain the reason for the thickening of the blood, but he emphasized the importance of the diet that could obviously influence the character of the blood. Since Boerhaave was of the opinion that fever made the blood thicker and therefore worsened the inflammation, which he saw as not only a local but also a general process and a major cause of disease, Boerhaave unlike Stahl favored medicines like quinine that lowered the fever. To prove his point, he made an experiment and placed a cat and a dog in an oven maintained at 63°C and sure enough both animals died within half an hour.

The Batavian Hippocrates, as he was sometimes called, certainly had an extremely broad education, including also the humanities and theology, and a remarkable grasp not only of medicine but also of botany and chemistry. In fact, he held chairs in all these three subjects at Leyden University. His international reputation was the main reason for the fame of its medical school, which attracted students from all over Europe, among them men such as Carolus Linnaeus and Albrecht von Haller, who would later become world famous in botany and physiology. Even the czar of all Russians, Peter the Great, attended Boerhaave's lectures during his stay in the Netherlands.

Boerhaave's greatness was above all as a teacher and perhaps his most lasting influence has been in chemistry. He started to give courses at the university in

Hermann Boerhaave (1668–1738).
Courtesy of Bonnierförlagen, Stockholm, Sweden.

Robert Boyle (1627–1691).
*Courtesy of the Royal Swedish Academy of Sciences,
Stockholm, Sweden.*

1701 and was appointed professor of chemistry in 1718; the chairs in medicine and botany he already held since 1709. His lectures in chemistry became extremely popular, not least because they were not entirely theoretical, also having a practical experimental aim. The very popularity of his chemical lectures among the students created something of a problem for Boerhaave. His listeners of course took extensive notes and some enterprising members of the audience went so far as to publish these notes as an unauthorized compendium under his name. Boerhaave was indignant when he discovered that this "sorry book" with its "false Notions, Absurdities and Barbarisms" enjoyed a resounding success among the medical students, so that in his own words "the wretched work was in everybody's hands."

In the end he was forced to write his own version and this is how his famous textbook *Elementa Chemiae* ("Elements of Chemistry") came to be published in 1732. He laid great weight on exact methods and experimental work rather than fanciful theories and introduced the use of thermometers and precision balances in chemical work. At the same time, he did not completely reject the claim by the alchemists to have transformed base metals into gold. His own extensive experiments with mercury, however, had not produced any gold even if they yielded mercury preparations of high purity as a spin-off. Boerhaave's wariness of indulging in speculative chemical theories could explain why in his textbook he does not even mention Stahl's celebrated phlogiston theory.

In his medical teaching, Boerhaave relied more on iatromechanical explanations than on the speculations of iatrochemistry, as expounded by Sylvius, that all illnesses are caused by a disturbance of the delicate balance between acid and alkali. Both van Helmont and Sylvius had stressed the parallel between fermentation to produce alcohol and vinegar, on one hand, and what we nowadays would call metabolism, on the other. On the whole Boerhaave rejected these theories, but he nevertheless considered chemistry and natural sciences in general to be of the first importance for the medical curriculum.

Fermentation was a central, if rather vague, concept in iatrochemistry. In the heyday of alchemy, it had been used to denote all sorts of chemical transformations brought about by the philosopher's stone, as well as a variety of elixirs and potions, in order to produce gold from base metals or to cure disease. Boerhaave instead used the term fermentation in a much more precise way to indicate a process by which vegetable materials were transformed into alcohol and later into vinegar. It is fair to say that here Boerhaave takes an important step towards making the half magic beliefs of iatrochemistry, originally inherited from alchemy, into a branch of natural science.

Corpuscles and Nitrous Air

A Great Experimentalist

The first earl of Cork, Richard Boyle, a typical Elizabethan figure renowned for his political acumen and notorious for his venality and greed, had no less than 14 children. The youngest of them, Robert (1627–1691), would achieve lasting fame as a philosopher, physicist and chemist, perhaps an unlikely career for a man with his background. He was sent to Eton and later on a tour of continental Europe accompanied by an elder brother. In addition to having received a thorough liberal education, he also early on came in contact with what was then referred to as "the new philosophy." Its foremost representative was the Italian physicist Galileo Galilei, whose famous *Dialogue of the Two Chief World Systems* Robert Boyle read while he visited Florence. Having returned to England, where the Civil War between Royalists and Parliamentarians had broken out, he was drawn to medicine and as a consequence began to take an interest in chemistry, the natural science closest to medicine. Under the influence of the leading minds of the time, Francis Bacon and René Descartes, he also took up the study of physics. Throughout his life, he insisted on the importance of using physical concepts in order to understand chemical problems, thus transforming the mysterious inheritance from alchemy into an exact natural science.

It was as a physicist that Boyle first achieved scientific fame through his studies of the properties of air. He used an improved version of the air pump first constructed by the German physicist Otto von Guericke, and worked with Robert Hooke (1635–1703), who had become his assistant. These studies led to the formulation of Boyle's law (the volume of a gas varies inversely with the pressure), as well as the discovery that air (in contrast to vacuum) is able to propagate sound. He also seems to have realized that air was necessary in order to sustain the life of an animal and the burning of a candle. In his most influential chemical publication, *The Sceptical Chymist* (1661), he refutes not only the elements of Aristotle (earth, air, fire and water) but also the ideas put forward by Paracelsus that mercury, sulfur and salt were the "principles" whose proportions in the living organism determined health and disease.

Instead of the metaphysical (not to say mystical) concept of "elements" he introduced what he called "corpuscles," a kind of elementary particles reminiscent of the atoms of the Greek philosophers Democritus and Epicurus. By thinking in terms of corpuscles and their movements one could understand such properties

John Mayow (1641–1679).
Courtesy of the Royal Society, London.

of matter as heat, fluidity and solidity. Furthermore, disparate corpuscles might come together and form groups leading to chemical reactions that give rise to new chemical compounds with properties different from the original reactants. To give an example from modern chemistry, oxygen and hydrogen can react to give water, a compound with entirely different properties than the reacting gases.

Perhaps Boyle's greatest importance has been as an untiring advocate of the paramount importance of the experiment. This was in contrast to such leading philosophers as Spinoza and Leibnitz, who held that logical reasoning was the only way to arrive at the truth; experiments could only confirm the conclusion reached through logical arguments. In consequence of his experimentally based empiricism, Boyle devoted much time and effort to working out a number of methods for the analysis of chemical compounds. He introduced, for instance, the use of indicators such as syrup of violets, which is turned red by acids, while alkalies turn the syrup green. In this way, he could show that the class of compounds, at this time referred to as "salts," in reality consisted also of acids and alkalies. The true salts were neither acids nor alkalies, i.e. they were neutral and gave no reaction with the indicator. His extensive contributions to the experimental arsenal of chemistry make him a pioneer of chemical methodology.

The Mouse and the Candle

John Mayow was born in Morval, Cornwall, in 1641, the scion of a well-established genteel family long resident at the manor house of Bray. He matriculated in 1658 at Wadham College, Oxford, and was elected a fellow of All Souls College there in 1660. He studied law at Oxford and became bachelor of common law in 1665 and doctor of law in 1670. There is no certain record of Mayow having obtained a medical degree, but he might have studied medicine while at Oxford. In any case, he is known to have practised medicine at Bath after having left Oxford in 1670. We know very little of his life from the time he finished his studies, except that he was proposed as a fellow of the Royal Society in 1678 by Robert Hooke and elected a member the same year. However, he was not destined to enjoy his membership in the Royal Society for long; he died in London the following year at the age of 38.

During his short scientific career Mayow published two books. The first, with the title *Tractus Duo*, was printed in Oxford in 1668 and contains studies of respiration as well as a treatise on rickets. His work on respiration is by far the most important part of the book, and here for the first time he argues that air contains something that he calls "nitrous air" and alternatively refers to as "particulae nitrosalinae" and "spiritus nitroaereus," which he considers to be necessary for life. These ideas he further expounded in a second book, published in 1674 under the title *Tractus Quinque Medico-Physici*. Here he shows that "spiritus nitroaereus" is taken up by the blood when it passes through the lungs. This is the reason for the difference in color of the bright red blood leaving the lungs, compared to the dark red of venous blood. Furthermore, he describes a number of experiments in which animals are breathing in a closed vessel like a cupping glass or a bell jar, resulting in the death of the animal when the nitrous air has been consumed. In the same way, a candle burning in such a closed bell jar went out when the spiritus nitroaereus was used up. He could also demonstrate that the breathing of the animal and the burning of the candle resulted in the reduction of the volume of the air in the vessel. Mayow also realized that when a metal like antimony was calcined, i.e. transformed into antimony oxide, it increased in weight because something (spiritus nitroaereus) was taken up from the air.

There are certain figures in the history of science that have something romantic about them and Mayow is a case in point. His early death, that cut short what appeared to be a very promising scientific career, of course appeals to our imagination. There is also his inspired vision of the true nature of respiration and combustion, which seemed far ahead of his time. With the advent of the romantic era at the beginning of the 19th century, there was a resurgence of interest in Mayow and his enigmatic writings more than a century ago that seemed to portend the great discoveries of Priestley, Scheele and Lavoisier. Perhaps he might even be seen as the true discoverer of oxygen? That is probably going a bit too far, even if he clearly realized the presence in the air of something that sustained both the life of a mouse breathing in a bell jar and the burning of a candle in such a jar.

On the other hand, he certainly did not produce oxygen in pure form as Priestley and Scheele did a hundred years later, nor did he have a clear concept of its chemical properties. Nevertheless, it seems fair to say that Mayow was the great pioneer of respiration and combustion in the 17th century and that he was probably more advanced in his thinking about these problems than were his contemporaries Boyle and Hooke. In any case, had he been taken seriously, he might well have prevented the rise of that unfortunate theory about "phlogiston," which was proposed at the end of the 17th century and would pervert chemical thinking for the following hundred years until the coming of Lavoisier and his antiphlogistic chemistry.

An Enigmatic Theory

The concept of "phlogiston" is often attributed to Georg Ernst Stahl (see above), but he had a predecessor in the German physician, chemist and adventurer Johann

Joachim Becher (1635–1682) whose thoughts about combustion are set forth in his treatise *Subterranean Physics* (1669). He discerned three elements that made up matter: air, water and earth. The element of earth was in its turn made up of the vitrifiable, the mercurial and the combustible earth. Becher assumed that when some inflammable material burned, combustible earth was liberated and given off to the surrounding air. This idea of something being lost to the air during combustion was taken over by Stahl, who called this disappearing matter "phlogiston." Whether this was actually a substance, possessed of a property such as weight, or should rather be regarded as the principle of combustibility, is not always obvious from his writings.

Stahl paid particular attention to the conversion of a metal into its oxide (at this time called a calx), which he saw as a kind of combustion that involved the loss of phlogiston, i.e. the metal had originally been composed of its calx plus phlogiston. The fact that the calx (the metal oxide) had increased in weight compared to the metal being oxidized, was a serious problem for the adherents of the theory and a number of more or less ingenious hypotheses were suggested to account for this. Strangely enough, no one seems to have come up with the straightforward explanation that the phlogiston theory was simply wrong. Instead, this upside-down way of reasoning, that is reminiscent of Alice in Wonderland, would warp the thinking about combustion and respiration for almost a century. It is significant that when Carl Wilhelm Scheele and Joseph Priestley independently produced oxygen in pure form in the 1770s, they were both completely captivated by the phlogiston theory. Priestley even called the new gas "dephlogisticated air" to emphasize the reverse relation between phlogiston and what was later called oxygen: the more oxygen in the air, the less phlogiston, and *vice versa*. A remarkable conclusion, one might think, but quite logical.

Metallurgy and Spas

Generally, iatromechanical ideas dominated medical thinking in the 17th and 18th centuries, but there was one field where the competing school, iatrochemistry, ruled supremely — spa therapy. The belief that the water of certain natural springs was useful for the treatment of a number of illnesses both when imbibed and when applied externally, was already widespread in the classical world. This time-honored therapy had a new lease on life with the advent of iatrochemistry with its emphasis on chemical explanations of medical problems. Springs in places like Spa and Aachen, whose waters were credited with marvelous therapeutic properties, became extremely popular and remarkable recoveries were reported when wealthy patients came from all over Europe to take the water cure there. A popular spa could mean a huge economic success and consequently there was great interest in the chemical analysis of spa water to establish its putative medical usefulness. This undoubtedly stimulated the development of analytical chemistry as applied to substances in solution, but there were also other even more economically important fields for chemical analysis.

After having successfully participated in the Thirty Years' War, Sweden clearly aspired to a position as a leading European power, an ambition that undoubtedly required ready money. The export of metals such as iron and copper was the chief source of cash for the Swedish government. This made chemistry, with its obvious importance for the development of metallurgy, the favorite science for the absolute sovereigns of the Carolingian period in that country. The leading representative of this, at least in Sweden newfangled science, was Urban Hiärne (1641–1724), chemist, physician and pioneer of spas in the Swedish countryside of which quite a few survive to this day. He was physician-in-ordinary to King Charles XI and chairman of the Royal Swedish Medical Board, but most importantly, he was head of the Royal Laboratorium Chymicum, which was housed in a newly renovated palace in Stockholm.

Hiärne was a good organizer and a competent analytical chemist. Under his direction, the Laboratorium Chymicum developed into one of the leading laboratories in Europe specializing in metallurgic problems. One of his early collaborators was the outstanding German chemist Johann Georg Gmelin, who later settled in Tübingen and became the progenitor of a famous dynasty of chemists. The tradition of metallurgy and mineralogy, which Hiärne had initiated, continued to dominate Swedish chemistry during the 18th century when it enjoyed a European fame. Its internationally best known representative was Torbern Bergman (1735–1784), professor of chemistry at Uppsala University. Swedish chemistry at the time had a down-to-earth character with minimal scope for originality and personal brilliance, but Bergman is the obvious exception to the rule.

An infant prodigy, after having matriculated at Uppsala University at the age of 17, he studied such disparate subjects as philosophy, mathematics, physics and entomology. At 23, he was appointed assistant professor of experimental physics and lecturer in mathematics. He was generally regarded as a young genius and when the professor of chemistry unexpectedly resigned his chair because of ill health, the crown prince (the future Gustavus III) who was chancellor of the University, saw to it that Bergman was appointed to the chair. In view of the fact that Gustavus was completely uninterested in science and that Bergman had never done any research in chemistry before, this was really a shot in the dark. Nevertheless, the unlikely choice of Bergman turned out to be a lucky hit indeed for Swedish chemistry. He continued the tradition of analytical chemistry and introduced the chemical analysis of minerals as the basis for their classification. Here he could build on the pioneering work of Axel Cronstedt, a leading chemist at the Laboratorium Chymicum.

As a theoretical chemist, he was a firm believer in the unfortunate phlogiston hypothesis, but he also realized the need for a new and systematic nomenclature and new chemical symbols that replaced the enigmatic signs for metals *et cetera*, that had at least in part been inherited from the alchemy of the Middle Ages. Unfortunately, both his nomenclature and his chemical symbols failed to be generally accepted, even if they in principle represented a considerable step forward and enabled the writing of equations of chemical reactions. It would take the authority of the dominant figure in the next generation of Swedish chemists, Jacob Berzelius, before the necessary reforms of the chemical language could be implemented. Perhaps Bergman's most important contribution to theoretical chemistry was his investigation of chemical affinity, i.e. the propensity of chemical substances to react with each other. His tables of affinity were generally hailed as a great advance, even if such tables had been published by several authors before. Bergman's writings were translated into all great European languages and he was even elected a foreign member of the French Academy of Sciences. Sadly, he was a victim of pulmonary tuberculosis and at the age of 49 he succumbed to his illness.

The Chemistry of Air

Aristotle had recognized air as one of the four elements, a concept that goes back even further to ancient civilizations like the Egyptian. It would take a long time before it was realized that the air that surrounds us is a mixture of gases that are chemically distinct from each other. Air might be important as a recipient of products given off during chemical reactions, but the idea that something in the air itself might actually participate in such reactions was alien to most chemists in the 17th century. For instance, it was believed that air did not take part in combustion, it was just there to receive the phlogiston given off by the burning material. However, during the latter half of the 18th century, this view of the air as a "passive" element would change dramatically.

Torbern Bergman (1735–1784).
*Courtesy of the Royal Swedish Academy of Sciences,
Stockholm, Sweden.*

Fixed Air

The Scottish physician and chemist Joseph Black (1728–1799) was born in Bordeaux, where his father was a wine merchant, but he was educated in Belfast and studied medicine at the University of Glasgow. Here he became a pupil and later assistant of the leading light at the medical faculty, William Cullen, who had inaugurated the study of chemistry as part of the medical curriculum. When Black transferred to Edinburgh, that could boast one of the best medical schools in Europe, he was not impressed by the teaching of chemistry there. In 1754 he submitted a thesis for his MD, in which he described experiments with the weakly alkaline substance "magnesia alba" (magnesium carbonate). He showed that this substance gave off a gas when heated and that the same gas was also formed when calcium carbonate was treated with strong acids. The gas was carbon dioxide, which had already been discovered by van Helmont in the previous century. Helmont had called it "gas sylvestre," but Black named the gas, that he had rediscovered, "fixed air" to signify that it had been present in the magnesia alba in a fixed form. He realized that in water solution, "fixed air" was weakly acidic, and that it was also produced during fermentation and respiration and the burning of charcoal. Unlike ordinary air it could not sustain the life of experimental animals, but he, nevertheless, concluded that it must be a normal ingredient of air. His later work, which led to the discovery of latent heat, was clearly of great importance, but does not concern us here.

Dephlogisticated Air

Of the scientists that were active in the elucidation of the chemical composition of air, Joseph Priestley (1733–1804) is decidedly the most original figure in terms of his broad interests in a number of rather disparate fields. He was a clergyman, who early in life demonstrated a very independent mind and professed opinions that involved him in conflicts both with the church authorities and, when he later took up teaching, with the trustees of the school where he was active. During his time as a pastor, he happened to live close to a brewery and he became interested in the gas (carbon dioxide) formed during fermentation, which had already been described first by van Helmont and later by Joseph Black. This led to a general interest in the chemistry of gases and in August 1774, he made an experiment where, using a burning glass, he heated mercuric oxide that was contained in a bell jar. He found that a previously unknown gas was formed, which was much more effective than ordinary air in sustaining the burning of a candle.

As the result of a famous meeting in Paris between Priestley and Lavoisier, which took place soon afterwards, the latter was made aware of the existence of

Joseph Black (1728–1799).
*Courtesy of the Royal Swedish Academy of Sciences,
Stockholm, Sweden.*

Joseph Priestley (1733–1804).
*Courtesy of the Royal Swedish Academy of Sciences,
Stockholm, Sweden.*

Carl Wilhelm Scheele (1742–1786).
Courtesy of the Royal Swedish Academy of Sciences,
Stockholm, Sweden.

the new gas, which he named "oxygen," while Priestley, being a confirmed believer in phlogiston, had called it "dephlogisticated air" (see above). Oxygen came to form the foundation of what was often referred to as Lavoisier's antiphlogistic chemistry, but Priestley remained an adherent of the phlogiston hypothesis to his dying day. Because of his religious views and the fact that he supported both the cause of the American colonists and later also the French Revolution, he became very unpopular in England and in the end was forced to emigrate to America, where he lived during the last ten years of his life.

A True Chemist

Carl Wilhelm Scheele (1742–1786) in every way conforms to the popular picture of a chemist always restlessly active in his laboratory, where he discovers and characterizes endless numbers of new compounds. He was born in Stralsund, the capital of Swedish Pomerania, in an originally well-to-do merchant family, but unfortunately his father went bankrupt when Carl Wilhelm was only a few years old. This was probably the reason why, at the age of 15, the gifted boy was apprenticed to a pharmacist in Gothenburg instead of receiving the university education that his talents merited. Luckily, his kindly employer allowed him to use the laboratory facilities of the pharmacy for his experiments. Here he spent all his free time, and he also avidly read chemical literature. He thus became a self-educated and very accomplished chemist and when in 1770, having worked both in Malmö and Stockholm, he came to the pharmacy Uppland's Arms in Uppsala, he may truly be said to have reached the peak of 18th century chemical knowledge both theoretically and experimentally. He made the acquaintance of Torbern Bergman and they became close friends and collaborated intimately. The five years that he spent here were probably his most productive scientifically and his working capacity was incredible. The number of important discoveries that he made during his fairly short life is indeed amazing.

He discovered the element chlorine by treating manganese superoxide with hydrochloric acid and he also managed to produce pure manganese. He prepared hydrofluoric acid from calcium fluoride and he was the first to purify arsenic acid, molybdic acid and tungstic acid. He also ventured into the realm of organic chemistry, rather unexplored at the time, and discovered for instance oxalic acid, citric acid, lactic acid and uric acid. He even attempted to characterize such extremely complicated biochemical substances as proteins and could show that they contained sulfur. His greatest discovery, however, must be said to be that of oxygen. Since he did not publish this until 1777 in his *Chemische Abhandlung von der Luft und dem Feuer* ("A Chemical Treatise of Air and Fire"), three years

Antoine-Laurent Lavoisier (1743–1794).
Painting by David. All rights reserved.
The Metropolitan Museum, New York.

after Priestley, he is seldom recognized as its discoverer. Nevertheless, when his copious laboratory journals were systematically examined in 1892, it became apparent that he had already in 1772 produced oxygen using several different methods. He also demonstrated the ability of oxygen (he called it vitriolic air) to sustain combustion. It is therefore fair to credit him with having, independent of Priestley and in fact a few years earlier, discovered this new element.

In 1775 he moved to the little town of Köping, where the local pharmacist had recently died and the young and not unattractive widow was prepared to transfer the privilege of her late husband as apothecary to Scheele. This meant that he now had a considerably better income and it would seem that his relations to the young widow were also rather amiable. Here he lived happy and content, in spite of attractive offers from such a great personage as Frederick II of Prussia to become professor of chemistry in Berlin. Scheele was not destined for a long life but his contributions to chemistry are solid and long-lived, and not encumbered with a lot of loose theoretical speculations but firmly grounded in untiring experimental work.

Combustion and Respiration

It would be an exaggeration to say that the phlogiston theory had ruled unchallenged since the end of the 17th century, but it is remarkable how many leading chemists that were prepared to rush to its defence and advance the most amazing hypotheses to explain such glaring inconsistencies as, for instance, the increase in weight when a metal was transformed into its calx (oxide) by the presumed loss of phlogiston. It was therefore an entirely new departure for chemistry when Antoine-Laurent Lavoisier (1743–1794) advanced his antiphlogistic concept of combustion and respiration. In fact, Lavoisier himself, who was never one to hide his light under a bushel, often talked of it as a revolution in chemistry.

Lavoisier was born in Paris, a city whose intellectual life at the time was dominated by the philosophers of the Enlightenment with names such as Voltaire and Diderot. His family, originally of humble, probably peasant stock, had moved up in society and was now fairly prosperous. His father was a solicitor at the Parlement of Paris (not a parliament but a kind of court) and had married the rather wealthy daughter of a well-to-do attorney. When Antoine was five years old his mother died, leaving him a considerable fortune. His father left Antoine and a younger sister in the house of his mother-in-law and here he spent his childhood in the loving care of his grandmother and an aunt, who never married and obviously adored the gifted and precocious boy. At the age of 11, he was admitted into the famous Collège Mazarin, founded in the 17th century by the politician and cardinal of that name. Here he received a broad education including mathematics and philosophy, languages and literature, as well as the sciences. Later he studied law, possibly to fulfil the wishes of his father, but his mind was early set on a scientific career, particularly in chemistry. To begin with, this was motivated by his interest in geology and mineralogy, which he felt should be firmly based on chemical analysis, and to the end of his life he continued to pursue these studies when time permitted. However, his boundless energy and amazing working capacity became more and more engaged in purely chemical problems.

Lavoisier had since an early age aimed at becoming a member of the prestigious French Academy of Sciences, and to this end he presented several scientific papers to the Academy. At the same time, influential relatives, in particular his father and aunt, pulled strings to further his ambitions. At 1768, at the age of 25, he was elected to the Academy as "adjoint chimiste surnuméraire" after having read a paper to the learned society in which he described a hydrometer of his own construction. At this time he also became a member of the "Ferme Générale," a private consortium that had acquired the right to collect certain taxes on behalf of the government. This would later prove to be a highly dangerous position during the period of the Terror that followed on the French Revolution. The same is also true of the title of nobility that his father had purchased shortly before his own death and that Lavoisier inherited in 1775. However, the threatening catastrophe still lay far in the future, while at the time everything looked rosy and promising for the young academy member. A further stroke of luck was his marriage in 1771 to Marie Paulze. The young bride, she was only 14 years old, immediately resolved to become her husband's collaborator in his scientific work. She began to study English, a language of which he had no knowledge, and took art lessons from the famous painter David, so that in the end she became a skilled draftsman and engraver who illustrated Lavoisier's books on chemistry. Marie also worked with him in the laboratory that he had created in the Paris Arsenal, where he became a scientific director in charge of the production of gunpowder. In fact, no scientist could possibly ask for a wife more devoted to his scientific work, and by all accounts their marriage was a model of harmonious bliss.

The problem that increasingly occupied his mind and that he worked on in his laboratory had to do with the chemical nature of air. He had been fascinated by an article in Diderot's *Encyclopédie* concerning the property of air (and generally of vapors) of being expansible or elastic as it was often called at this time. The anonymous article, probably written by the well-known economist, philosopher and politician Robert Jacques Turgot, suggested that matter could exist either as a solid, a liquid or a gas, depending on how much "fire" it is combined with. This led Lavoisier to believe that air might exist in two forms: free, i.e. as a gas, or fixed in combination with a number of substances. This is of course reminiscent of Joseph Black and his fixed air, the carbon dioxide that could be liberated from calcium carbonate by treatment with acids and that also was a normal component of air. However, it is not clear to what extent Lavoisier at this time (1772) was aware of Black's discovery (see above).

Regarding another component of air, what we now call oxygen, it had been known even in the heyday of the phlogiston theory, that metals like lead and tin gained in weight when they were transformed into a calx (oxide) and a number of fanciful hypotheses had been proposed to account for this phenomenon. Lavoisier, however, became increasingly convinced that the weight increase of the metal was caused by it having taken up something from the air when it formed a calx. Furthermore, in 1772 a Paris pharmacist, Pierre Mitouard, reported that air seemed to be consumed when phosphorous was burned with the formation of phosphoric acid. Lavoisier now carefully and systematically investigated the burning of phosphorous and sulfur and could definitely show that air was consumed in these reactions and that the corresponding acids formed (phosphoric and sulfuric acid) weighed more than the starting material, phosphorous and sulfur. It was obvious that air, or some component of air, had been taken up.

These results raised the question of whether the calx of a metal could liberate the component that had been taken up from the air during its formation, if it were heated to a sufficiently high temperature. It had already been demonstrated in 1774 by two French chemists that the red calx of mercury (mercuric oxide) could be reduced to metallic mercury when heated. It is now that Lavoisier becomes aware of Joseph Priestley and his dephlogisticated air during a visit by the latter to Paris. As we know, Priestley had produced the gas by heating mercuric oxide, but to begin with he mistakenly identified it as nitric oxide, a gas that he had previously discovered. However, he soon rectified this mistake even if he continued to believe that what he had produced was air free of phlogiston. Furthermore, Carl Wilhelm Scheele had written in 1774 to Lavoisier and reported that a gas, which very effectively sustained combustion, could be obtained by heating silver carbonate. One could say that everything was now ready for Lavoisier's great discovery of the role of oxygen in combustion and respiration. He had originally described the gas that he obtained by heating mercuric oxide as "highly pure atmospheric air," but he now denoted it "the purest part of the air," or alternatively, "eminently respirable air," indicating that he saw it as something that constituted a normal part of air. Lavoisier also performed experiments in which he exhausted the ability of air to oxidize mercury to a calx. He could then demonstrate that the remaining gas could not sustain either the burning of a candle or respiration, nor did it form a precipitate with limewater, i.e. it was not fixed air. This gas was nitrogen and had previously been obtained in similar experiments by Scheele and Priestley. Thus, the air that surrounds us is made up of three components: "eminently respirable air," nitrogen (Lavoisier called it "mofette atmosphérique"), and Black's fixed air (carbon dioxide).

We must now turn our attention to the group of substances, the acids, which had been distinguished by Robert Boyle among what was then loosely called "salts," because of the ability of acids to react with certain indicators like syrup of violets. During the 18th century, it was realized that salts could be formed when alkaline substances were neutralized by acids or when metals came into contact with acids. A number of new acids were also discovered, and Carl Wilhelm Scheele was particularly active here. At the same time, the problem of what properties constituted an acid still remained unresolved. It was widely believed that there existed what might be called a kind of progenitor from which all other acids could be derived and several candidates for this primeval acid were suggested, for instance sulfuric acid and carbonic acid. Another possibility was the "oily acid" (acidum pingue) advocated by the German chemist J.F. Meyer, which strangely enough was supposed to confer the properties of a base in some instances, and of an acid in other cases. Even Lavoisier seems for a while to have been impressed by Meyer's acidum pingue, but his experiments with the burning of phosphorous and sulfur to form the corresponding acids made him change his mind. In his *Opuscules Physiques et Chimiques* (1774), he wrote that phosphoric acid was "in part composed of air, or at least of an elastic substance contained in the air." Two years later, in a paper that he read to the Academy, Lavoisier concluded that *all* acids were in great part made up of air and that it was what Priestley had called dephlogisticated air that formed part of the acids. As an example he mentioned the formation of mercuric nitrate from mercury and nitric acid. When heated, the mercuric nitrate was converted to mercuric oxide from which on further heating "eminently respirable air" was liberated. Clearly, it must have originally formed part of nitric acid. In 1781, he published a paper about the general nature of acids in which he concluded that "eminently respirable air" was really the acidifying principle itself and he proposed to call it "principe oxigene", i.e. the begetter of acids. When the *Nomenclature Chimique* was published in 1787, the new gas was called "oxygène" (oxygen).

Certainly, Lavoisier was wrong here, oxygen is not the begetter of acids, at least not of all of them. There was, for instance, hydrochloric acid and hydrofluoric acid, recently discovered by Scheele, and already at this time it was realized that they did not seem to contain oxygen. Lavoisier explained this exception to his rule of thumb by blaming faulty analyses, which just illustrates how stubbornly even the greatest scientists sometimes cling to their favorite hypotheses. On the other hand, very many of the acids known at the time, for instance nitric acid, phosphoric acid and sulfuric acid, not to mention all the organic acids that Scheele had discovered, actually contained oxygen. Therefore, Lavoisier may well be excused for believing that oxygen was the begetter of acids, even if his unwarranted generalization were to cause some confusion before his mistake was eventually rectified.

In the early 1780s, balloons that could fly by the hot air principle were all the rage. In order to keep them in the air the crew had to use fires fuelled by sundry light combustible materials like straw *et cetera*. These hot air "aerostats" were of course far from reliable in terms of their ability to remain airborne for any prolonged time, even if pioneers like the Montgolfier brothers had demonstrated that unmanned hot air balloons could actually fly and Pilatre de Rozier together with the Marquis d'Arlandes had gone up in such a balloon in the first manned flight over Paris in 1783. What was obviously required in order to make these flights more reliable was a balloon filled with a gas lighter than air. Already at this time Jacques Charles, a French physics teacher, had used hydrogen to fill an unmanned balloon that he released over Paris. Soon afterwards Charles and a companion took off in a hydrogen balloon that rose to a height of 300 fathoms and whose altitude could be regulated by using bags of sand as ballast and a vent to release hydrogen from the balloon. The only problem was really to produce hydrogen in sufficient quantity.

Tradition has it that Paracelsus was the first to obtain hydrogen when he immersed metal in acid, which led to the formation of an inflammable gas. However, he did not distinguish this gas from other inflammable gases such as hydrocarbons. It was not until 1766 that Henry Cavendish characterized what he called "inflammable air" as distinct from other combustible gases. In 1781, he could demonstrate that water was formed when the inflammable air burned in the presence of dephlogisticated air. At first, members of the French Academy of Sciences doubted this report, but when the experiment was repeated on a larger scale by Lavoisier in the presence of Cavendish's assistent, Charles Blagden, water was undoubtedly formed. Thus, the reaction between hydrogen and oxygen to form water was definitely established, even if it had not been possible to measure the exact amount of water produced. Obviously, water was not an element as Aristotle had believed, and the inflammable air of Cavendish was soon afterwards renamed "hydrogen," meaning "the begetter of water."

Lavoisier had become a member of the standing committee appointed by the Academy to improve the construction of balloons, including methods for the production of suitable gases lighter than air. Hydrogen was clearly the best choice here and Lavoisier decided to use the method of exposing water to red hot iron that had previously been found to decompose water with the formation of hydrogen. The fact that hydrogen could be formed by the decomposition of water was not only a technically important method for the filling of flying balloons with a gas lighter than air. Together with the demonstration that hydrogen burned in oxygen to produce water, this was fundamental for Lavoisier's basic thinking about his antiphlogistic chemistry. He had made his general ideas about combustion known, albeit cautiously, already in 1777 at the same time that he rejected the phlogiston theory as untenable, (*Mémoire sur la Combustion en General*). In the period 1782/83 he collaborated closely with Pierre Laplace, a young collegue in the Academy, using an ice calorimeter (as suggested by Laplace) to determine the heat given off in chemical reactions by measuring the amount of water formed by the melting ice. Among the calorimetric experiments performed by Lavoisier and Laplace was also the determination of the heat produced by a guinea pig placed for several hours within their apparatus. This can be said to represent Lavoisier's first essay in physiological chemistry.

In 1786, Lavoisier published his definitive refutation of the phlogiston hypothesis: *Réflexions sur le Phlogistique*. This brought him in conflict with not only Joseph Priestley, who remained a steadfast adherent of phlogiston to the end of his life, but also a number of influential French scientists. On the other hand, such leading chemists as Claude Berthollet and Antoine de Fourcroy were converted to the new antiphlogistic ideas. In particular, Fourcroy's immensely popular textbook *Elémens d'Histoire Naturelle et de Chimie* became an important vehicle for the spreading of Lavoisier's message. However, the most important account of the new and revolutionary chemistry is undoubtedly Lavoisier's own classic work, *Traité Élémentaire de Chimie*, published in 1789, the same year as the outbreak of the French Revolution; perhaps a significant coincidence in time. The prefatory chapter of this book, *Discours Préliminaire*, is probably the best known of all that Lavoisier ever wrote. In his Traité and the prefatory *Discours* he is much preoccupied with the general principles derived from the basic concepts of the Enlightenment. At the same time, he wrestles with the eternal problem of the elements, what he calls the "substances simples," and how they should be defined. He obviously tries to reduce the number of elements to something more manageable and in order to achieve this he lays down certain rules. The elements must not only be indivisible, i.e. they cannot be further decomposed, but must also be widely distributed in nature. Consequently, precious metals like gold, which are only found in small quantities, cannot be elements. However, from a modern point of view, the most interesting part of his *Traité* deals with what we should call biochemistry. Take for instance his work on the "vinous fermentation" of sugar to alcohol. Under the influence of ferment, the elements of oxygen, hydrogen and carbon, present in the sugar, appear also in the products, i.e. alcohol and carbon dioxide, that result from the fermentation process. Nothing is created or lost in this reaction, says Lavoisier, it is just a question of alterations and modifications.

Lavoisier had long been interested in the problem of respiration. Already in 1777, he maintained that it was the oxygen in the air, which was involved in the respiration. He declared that this process was similar to the combustion of carbon and thus must account for the production of animal heat. In the famous experiment with the guinea pig in the calorimeter, he and Laplace had found that the heat produced by the animal was roughly comparable to the heat obtained by burning a piece of carbon with the formation of the same amount of carbon dioxide as that given off by the guinea pig. However, they noted that more heat was produced by the guinea pig, for a certain amount of carbon dioxide formed, than was given off by the burning coal. In the same vein, the animal produced less carbon dioxide than should have been expected from the amount of oxygen consumed. This led Lavoisier to assume that some of the oxygen was consumed in the formation of water from hydrogen. In further experiments together with Armand Seguin, they demonstrated that the amount of oxygen consumed by the animal increased with temperature and exercise. They furthermore, erroneously, concluded that the production of animal heat takes place specifically in the lungs.

While these experiments, which represent the first pioneering investigations of energy metabolism, were underway in the early 1790s, the French Revolution proceeded inexorably on its fatal road towards the unrestrained Terror that culminated in 1793/94. To begin with, Lavoisier's scientific career and steady rise in the Academy to the top rank of "pensionaire" seemed unaffected by the political convulsions in Revolutionary France, but this would soon prove to be an illusion. Lavoisier's father had shortly before his death purchased a title of nobility for himself and in 1775 his son inherited the title. At the time, this inheritance seemed innocent enough, but in the end it would turn out to be disastrous. Another fatal circumstance was Lavoisier's position as a member of the Ferme Générale (farmers-general). This was enough to make him a target of vehement attacks by radical journalists like Jean-Paul Marat. In 1793 he was removed from his position at the Arsenal, where he had been in charge of the production of gunpowder, and the same year also witnessed the suppression of the Academy of Sciences. The National Assembly had already abolished the hated Ferme Générale and in December, 1793, Lavoisier and his father-in-law were arrested together with all the other farmers-general. On 8 May 1794, they were brought before the Revolutionary Tribunal, summarily convicted and executed the same day with the aid of the guillotine, that convenient instrument for the disposal of the enemies of Freedom and Liberty. Of this mindless crime, committed in the name of Freedom and Equality, the great mathematician Joseph-Louis Lagrange sadly remarked: "It required only a moment to sever that head, and perhaps a century will not suffice to produce another like it."

If the 17th century can be said to usher in modern physics, with Newton as its dominating name, the 18th century is the period when chemistry definitely makes its debut on the scientific scene. There are several important actors here but none that can compare with Lavoisier and his introduction of the antiphlogistic chemistry. His elucidation of the true nature of combustion and the role of oxygen in this process is certainly one of the most revolutionary events in the history of chemistry. At the same time it can be seen as the birth of biochemistry.

A GOLDEN AGE
of CHEMISTRY

Proportions in Chemistry

A Pupil of Kant

After having graduated from the Gymnasium in Hirschberg, then a small town in Prussia, Jeremias Benjamin Richter (1762–1807) joined the engineering corps of the Prussian army, but he obviously found military service intellectually less than stimulating. In any case, after seven years he left the army in order to study mathematics and philosophy at the University of Königsberg, where the great German philosopher Immanuel Kant was the leading light and professor of logic and metaphysics. In 1789, Richter was awarded the doctorate based on a dissertation about the use of mathematics in chemistry. His thesis can be said to have been in the spirit of his idol Kant, in the sense that Kant always maintained that all true science is applied mathematics. Consequently, Kant saw chemistry as what he called a systematic art rather than a real science, but here Richter obviously was of a different opinion. Throughout his lamentably short scientific career, he always maintained that all chemical processes must be understood based on the principles of mathematics. This firm conviction would eventually lead him to the formulation of the important concept of stoichiometry.

After his dissertation he supported himself as a chemist and maker of aerometers (instruments for the determination of the density of air and other gases), but in 1798 he obtained a position as chemist at the Royal Porcelain Works in Berlin where he stayed until his death from tuberculosis at the age of 45. He never held an academic position but his scientific productivity was, nevertheless, considerable including three volumes entitled *Anfangsgründe der Stöchyometrie oder Messkunst Chymischer Elemente*, published in 1792–1794 ("Introduction to Stoichiometry or the Measurement of Chemical Elements"). In the first volume of his magnum opus, he defines stoichiometry as "the science of measuring the quantitative proportions or mass ratios in which chemical elements stand one to another." In this spirit he devoted his life to the search for the laws that govern the proportions in which the reactants in chemical reactions are combined. His ideas of equivalent proportions would turn out to be fundamental in the new chemistry that developed at the beginning of the 19th century.

An Apothecary and Analyst

When Joseph Louis Proust (1754–1826) took up pharmaceutical studies, he followed both a natural inclination for analytical chemistry and the family tradition, his father having a pharmacy in Anger where he got his first training.

Proust completed his studies in Paris but at the age of 24 he moved to Spain, the country where he would spend most of his professional career. However, his first visit to Spain was fairly short and in 1780 he was back in Paris, where he worked with Pilatre de Rozier and Jacques Charles (see above) on aerostatic experiments. These ventures culminated in the ascent of Proust and Pilatre in a balloon filled with hydrogen over Versaille in June 1784, watched by a public that included the Kings of France and Sweden and their courtiers.

In 1786 Proust was offered a position as professor of chemistry in Madrid, where he stayed for two years, and then in Segovia at the Royal Artillery College there. He remained in Segovia until 1799 when he moved to a superbly equipped laboratory in Madrid. Working conditions there seem to have been excellent, but in 1806 he had to return to France as a consequence of political and military conflicts between France and Spain during the Napoleonic Wars. He ended his career in Anger, the place of his birth, where he took over the family pharmacy after his brother, who had to retire because of ill health.

Proust's great scientific discovery, the definite proportions in which chemical elements occur in different compounds, owes much to his skill as an analyst. His analyses of iron oxides, that he published in 1794, convinced him that these compounds were of two different kinds; containing either 27% or 48% of oxygen, while contrary to what had previously been believed there were no intermediary stages between these two extremes. In a series of papers, he presented evidence that this was true also for many other metal oxides and that when such a metal oxide, with a low percentage of oxygen, was transformed into one with a higher oxygen content, this was always accomplished in one, single step. There was never any formation of oxides with an intermediary oxygen percentage. All results that seemed to indicate such intermediary stages, he explained as mixtures of what he called "maximum and minimum oxides." It was just a question of the investigator not having been able to separate the minimum and maximum oxides from each other. Consequently, he had been analyzing a mixture, not a pure oxide.

In the meantime, the great French chemist Claude Louis Berthollet (1748–1822) had by 1801 arrived at the conclusion that chemical elements could react in different proportions so that the resulting compound could represent a continuum of proportions of its constituent elements (atoms). This led to a conflict between him and Proust, who in 1804 published results on metallic oxides, which indicated that Berthollet's results were caused by this investigator having analyzed mixtures of two oxides, or alternatively, of the metal and one of its oxides. The dispute continued for several years and it would seem that the two combatants held to their views until the end of their lives. Nevertheless, there can be no doubt that, by introducing the principle of definite proportions of the elements in a chemical compound, Proust has made a major contribution to modern chemistry.

Atoms and Molecules

The Ingenious Quaker

The self-educated, completely dedicated amateur scientist has played a major role in the history of English science and John Dalton (1766–1844) is a case in point. He came of a lower middle-class Quaker family in Cumberland without any intellectual or economic background that might have facilitated a scientific career for the gifted boy. In the local Quaker school that he attended, Dalton attracted the attention not only of his teachers but also of a prominent member of the Quaker Society of Friends, the naturalist Elihu Robinson. This was probably why, at the age of 15, the boy was rescued from being put to work as a laborer by an invitation to become an assistant in a Kendal boarding school. Here he had access to the school's rather impressive library in the natural sciences and its fairly well equipped laboratory. There was also a steady flow of lectures by well-known naturalists that he could attend. Of particular importance for Dalton's scientific thinking was his contact with the Kendal Quaker and natural philosopher, John Gough, who taught him mathematics, meteorology and botany.

In 1785, John Dalton and his brother Jonathan took over the Kendal school upon the retirement of its former principal. John eked out his incomes by giving public lecture courses in such subjects as mechanics, optics, astronomy, etc. His educational talents must have been appreciated by the Society of Friends, because in 1792 he was appointed professor of mathematics and natural philosophy at the New College, which the Society had recently established in Manchester. This of course meant a great opportunity for him; he seemed well pleased with his new position, even if his teaching duties felt somewhat overwhelming. However, in 1800 Dalton resigned his professorship and in the same year he opened a private "Mathematical Academy" in Manchester where he taught mathematics, experimental philosophy and chemistry. The Academy turned out to be an immediate success and during the rest of his life he never held a university position but supported himself adequately as a private teacher of mathematics and natural sciences.

His period as professor at the Manchester New College had by no means been barren scientifically. In 1794, having become a member of the Manchester Literary and Philosophical Society, he read his first scientific paper there entitled "Extraordinary Facts Relating to the Vision of Colors, with Observations." Another major interest was meteorology and the construction of such instruments as barometers, thermometers and hygrometers. It was only natural that his studies in meteorology should include also the properties of water vapor and other gases, which make up the atmosphere. In his *Meteorological Observations and Essays*

John Dalton (1766–1844).
Courtesy of the Royal Swedish Academy of Sciences, Stockholm, Sweden.

(1793), he advanced the idea that each gas acts as an independent entity when present in a mixture of gases. In 1801, he was ready to publish his "New Theory of the Constitution of Mixed Aeriform Fluids, and Particularly of the Atmosphere" in William Nicholson's *Journal of Natural Philosophy, Chemistry and the Arts*. Here, he clearly states that in a mixture of two different gases, denoted A and B, respectively, the particles of A do not repel those of B, as they do one another. This early rendering of the law of the partial pressure of gases, was further elaborated in *The Absorption of Gases by Water and Other Liquids* (1803). These new and controversial ideas about gaseous mixtures and their properties were criticised and even ridiculed by many contemporary chemists, for instance Berthollet, but Dalton undauntedly defended his theories.

In the meantime, Dalton's interest was beginning to shift from the physical properties of gases to fundamental problems of chemistry. For instance, "why does not water admit its bulk of every kind of gas alike?" He ultimately became convinced that this was because different gases are made up of particles of different weights. In the same way, chemical reactions involve the combination of particles that differ in weight. Dalton's atomic theory is outlined in his *System of Chemistry* (1807) and *New System of Chemical Philosophy* (1808). Fundamental to his theory was the concept of chemical elements, the conservation of mass in chemical reactions (Lavoisier) and the definite proportions of elements in pure chemical compounds (Proust). Dalton now postulated that matter is made up of atoms, and that atoms of the same element have the same weight and are identical in all other properties. On the other hand, atoms of different elements, for instance hydrogen and oxygen, have different weights and differ also in other respects. Furthermore, atoms are indestructible and chemical reactions only imply that atoms are being rearranged, although they themselves remain unchanged.

Obviously, many of the basic concepts of Dalton's atomic theory had been around for quite some time. What was new and important was the concreteness of his thinking and the fact that his theory was precise enough to make predictions that could be tested experimentally. To begin with, the implications of his atomic theory for chemistry were not generally realized but this would change over the years, not least through his own efforts as an author and a lecturer. He was a leading member of the Manchester Literary and Philosophical Society and finally became its president in 1817. Five years later, he was elected a fellow of the Royal Society. Dalton never married and lived with a clerical friend, the Rev. W. Johns, at the same address in Manchester until his death. Apart from his membership in learned societies he seems to have been something of a recluse whose days were filled with laboratory work and lectures. In any case, the contributions to modern chemistry of this somewhat eccentric bachelor can hardly be exaggerated.

Humphry Davy (1778–1829).
*Courtesy of the Royal Swedish Academy of Sciences,
Stockholm, Sweden.*

An Astonishing Career

Humphry Davy (1778–1829) was born of yeoman stock in Penzance, Cornwall. He seems to have been an exuberant, imaginative boy, sociable and generally popular, in every way the opposite of John Dalton, the other leading English chemist at the turn of the century. Certainly, there was nothing in his background and his rather haphazard schooling that in any way portended his future illustrious career as a scientist. Later in life he would claim that his sketchy education had given him ample time for self-education. In any case, he seems to have read a good deal of both science and philosophy as a young man, including Lavoisier's *Traité Élémentaire de Chimie* in the original French. Based on this wide but somewhat unsystematic reading, Davy embarked on a career as a chemist and in 1798 he was appointed chemical superintendent of Thomas Beddoes' Medical Pneumatic institution, an establishment that was dedicated to investigating the possible therapeutic uses of various gases. Davy threw himself into this field with characteristic energy and enthusiasm. The results of his efforts, which included some rather hazardous experiments, where he inhaled poisonous gases to ascertain their physiological effects, were published in 1800 and he was invited to lecture in chemistry at the Royal Institution in London. Two years later he was appointed professor of chemistry there.

Humphry Davy was not only an exceptionally gifted scientist, he also had remarkable social talents, and it is typical of him that already as a young man his career was sponsored by such luminaries in British science as Sir Joseph Banks, Henry Cavendish and Benjamin Thompson (Count von Rumford). He was also a great communicator, who from an early age made a name for himself in the popularization of science. At the same time, he had an intuition in scientific matters that allowed him to select problems that would prove to be fruitful and important. His work on electrolysis using Alessandro Volta's newly invented pile is a good example of this. He was convinced that in electrolysis the current induced the separation of compounds into their elementary components rather than the synthesis of new substances, as many scientists believed at the time.

In 1807, Davy made an important discovery when he subjected slightly moistened potash (potassium carbonate) to electrolysis. He noticed that a silvery matter was deposited at the negative pole, while at the positive pole oxygen was liberated. Davy surmised that the silvery matter observed at the negative pole was of a metallic nature and called it potassium. In similar experiments with sodium hydroxide he also characterized the metal sodium. He then went on to electrolyze the so-called alkaline earths, which led to the isolation of magnesium, calcium, strontium and barium. Davy announced his remarkable discoveries in a series of Bakerian Lectures (he gave all in all no less than five such lectures) and his fame

Jacob Berzelius (1779–1848).
Courtesy of the Royal Swedish Academy of Sciences,
Stockholm, Sweden.

both in England and internationally continued to grow. His work with the alkaline earths showed them to be oxides, while Lavoisier had maintained that oxygen was a necessary part of all acids. However, Davy's investigation of what was then called "muriatic acid" (hydrochloric acid) convinced him that this acid did not contain oxygen. Furthermore, Lavoisier had claimed that what we today call chlorine was really a higher oxide of muriatic acid, which he called "oxymuriatic acid." Davy could show, however, that muriatic acid was in reality hydrochloric acid and did not contain any oxygen; by the same argument oxymuriatic acid was not an acid at all and did not contain any oxygen. It was instead the element chlorine.

Recognition and honors came early to Davy. Already in 1803 he was elected a fellow of the Royal Society. In 1812 he was knighted and 1813–1815 he made a European tour and in spite of the war with Napoleon, which was still ongoing, he was able to meet a number of outstanding French colleagues and even to perform experiments on a substance akin to chlorine, later called iodine. He also collected the medal awarded to him by the Institut de France for his discoveries on electrolysis. In 1818 he was made a baronet and in 1820, on the death of Sir Joseph Banks, he succeeded him as president of the Royal Society. This can be seen as the peak of his career, but in the 1820s his health began to fail and in 1826 he suffered his first stroke. His last years were spent on restless journeys around Europe, but he never fully recovered and in 1829, a fatal stroke ended his life in Geneva.

Atomic Weights and Symbols

An Unruly Schoolboy

Today, in our secular society, it is perhaps difficult to fully appreciate the importance of the vicarage in rural Europe as a nursery for future scientists and the bearer of an intellectual tradition in the rather austere cultural climate that prevailed there two centuries ago. At least, this was certainly true for Sweden, situated as it was in the outskirts of Europe, far from the cultural highroads.

Jacob Berzelius (1779–1848) came from a family of clergymen and his father, who died when the boy was only four years old, was also ordained even if he worked as a teacher. His mother soon remarried the Rev. Anders Ekmarck, a widower with five children of his own. The boy grew up in a rural vicarage in the south of Sweden, not far from lake Vättern, but when he was eight years old he lost his mother to the same illness that had carried off his father, pulmonary tuberculosis. His stepfather, whom he portrays a very positive picture of in his autobiography, was now left with no fewer than eight children to care for and one can understand if he found his responsibilities somewhat overwhelming. As long as his former wife's younger sister took care of the household the domestic situation at the vicarage was manageable, but when she married a couple of years later, the crisis became acute. The vicar's attempt to solve his problems by marrying the widow of a clergyman did not turn out well, at least not from his stepson's point of view. In the end, Jacob had to move to a maternal uncle with seven children and an alcoholic wife. The time he spent there was obviously miserable and it must have been a relief for him when he was admitted to the Gymnasium in Linköping. Even so, his schooldays were certainly not without problems.

Like most other schools at the time, the Gymnasium in Linköping emphasized the humanities and classical languages, while natural sciences played a much smaller role in the curriculum. The mindless swatting of Latin and Greek was not to the liking of the young Jacob Berzelius, who much preferred excursions in the rural surroundings where he could study Nature armed with a shotgun. These extracurricular activities were brought to the attention of the headmaster, who was deeply shocked and immediately forbade them. Jacob did not take this prohibition of his hunting tours so seriously and a couple of days later he was out with his gun again. Unfortunately it so happened that he ran into the headmaster a second time and now the consequences threatened to be an unmitigated catastrophe. Next day at morning prayers the headmaster announced that Jacob should be punished with the birching rod and this was obviously meant to take place immediately. Fortunately, the delinquent had taken a few hours leave from school in the unusually fine spring weather and was nowhere to be found. In the meantime his friends warned him of what was in the wind. He was naturally appalled at the prospect of having his bottom whipped in public, but what was even worse, he might be expelled from the school as an additional punishment. Luckily, Jacob had a friend and patron among the teachers, who had not been consulted about the disastrous verdict. He became both insulted and indignant and appealed to the bishop, who was the inspector of the school. Through the intervention of that August personality, Jacob's backside was saved from the birching rod and nothing came of the threatening expulsion. However, in the end the vindictive headmaster had his revenge and Berzelius' leaving certificate from the Gymnasium was something less than satisfactory. According to Berzelius' autobiography, the summing up of his certificate described him as a young man of good natural talents, but with unsatisfactory morals and a questionable future. He may have exaggerated a bit here (in the original Latin the certificate was not quite so catastrophic) but it is, nevertheless, the most notorious and least perceptive character ever given in a Swedish school.

A Leading Chemist

Having escaped from the Gymnasium, Berzelius decided to use the small inheritance from his mother to help him through medical school at Uppsala University. It would appear that his intent was really to study chemistry rather than to pursue a medical career. In fact, using medicine to secure a living was nothing unusual for leading scientists at the time; Carolus Linnaeus had been a professor of medicine in Uppsala from 1741 until his death in 1778. Berzelius had tried to master the theory of chemistry by reading books like de Fourcroy's *Philosophie Chimique* and Christopher Girtaner's *Introduction to Antiphlogistic Chemistry*. After these preparations he called on the professor of chemistry and asked permission to do experimental work. This was granted, albeit very reluctantly, but Berzelius soon found that the laboratory to which he had been admitted was utterly primitive and devoid of practically all equipment. Faced with this lack of encouragement on the part of his professor, Berzelius typically adequately equipped a simple laboratory of his own in his student quarters and here he managed to produce pure oxygen. Perhaps not a very original endeavor, but the result was, nevertheless, highly encouraging for the future chemist.

Taking the waters at various spas had a long tradition in Sweden, and as a young doctor Berzelius became associated with the famous spa in Medevi. He analyzed its water and also became acquainted with a newfangled therapeutic method, the use of electricity from the pile of Volta to treat sundry illnesses among the guests that patronized the spa. His results with this treatment formed the basis for the thesis, which he submitted to the University of Uppsala in fulfilment of the requirements for his MD in 1802. There was really nothing about this dissertation that in any way portended the eminent position that the young Dr. Berzelius would achieve in the history of chemistry. However, he obviously had the gift of friendship and one of his friends was the wealthy mill owner Wilhelm Hisinger, whose acquaintance he had made at the Galvanic Society in Stockholm. Their common interest in electricity and its influence on salt solutions led to a lifelong friendship as well as a series of important experiments.

Together with Hisinger and other collaborators, Berzelius independently obtained the same results as Humphry Davy in his electrolysis experiments. He was offered an unsalaried position as lecturer at what was a few years later to become the Karolinska Institutet. Here he was entrusted with the task of teaching the future military surgeons the elements of what we today call biochemistry. He soon found that there was an urgent need for a textbook in this rather newfangled subject and in 1806 he published the first volume of his classical book *Föreläsningar i Djurkemien* ("Lectures in Animal Chemistry"), one of the very earliest textbooks in biochemistry. It is interesting to note the decisive way in which the young author rejects the natural philosophy that dominated the thinking in biomedicine in Continental Europe at the time.

In 1807, Berzelius was appointed professor of pharmacy and chemistry at the medical school in Stockholm, whose main function was to turn out military surgeons. This, for the first time in his life, made him economically secure. He had previously supported himself as a physician for the poor. Even if he conscientiously toiled with his animal chemistry, mindful of his duties at the medical school, Berzelius was increasingly attracted by inorganic chemistry, which was much more amenable to analysis that could yield definitive and reproducible results. His determination of atomic weights is a good example both of his ability to obtain chemical compounds in pure form and his analytical skill. The fundamental principles were already known, but it required Berzelius' remarkable experimental talent and inexhaustible capacity for work to achieve the goal. The values of atomic weights, which he determined using the very primitive technical means available at the time, are amazingly accurate and agree closely with modern figures.

A Chemical Language

A modern chemist without any knowledge of the history of chemistry, would be hopelessly confused if he were confronted with a paper written by Lavoisier, Scheele or Dalton. It would be full of enigmatic symbols, not unlike the hieroglyphics used in ancient Egypt and to make things worse, different authors used their own personal symbols totally unrelated to each other. To a man like Berzelius, with his demands for a logical and systematic presentation of chemical facts, such disorder must have been intensely irritating. He decided to do something about this mess. He had certainly not been enamored by the Latin swatting of his schooldays but he, nevertheless, realized that an international system of symbols, made up of letters, must take as its starting point the only truly international language — Latin. When he introduced his chemical symbols in 1811 in the *Journal de Physique*, he used the first letter in the Latin name of an element, for instance C for carbon (carbo) while for copper he had to introduce an additional letter Cu (cuprum). He also used index numbers to designate the proportions of atoms making up a molecule, for instance H_2O for water. The order and method in the presentation of chemical data that Berzelius' system made possible, was decisive for the rapid development that now followed and that would have been impossible without a generally accepted, common language of symbols.

One could perhaps say that Berzelius' greatness as a scientist was his ability to create order out of confusion, rather than to come up with new ideas and theories, but there is one very important exception to this generalization. From 1821, he had published his celebrated and very influential *Årsberättelser* ("Annual Reports") and here in 1835 he introduced the idea of catalysis, one of the most fundamental concepts in chemistry. He gives examples of catalysis in inorganic chemistry and draws a parallel to Kirchhoff's discovery that strong acids can bring about the formation of glucose from starch without themselves being consumed in the reaction. He then goes on to define catalysis and catalyst and he emphasizes that the catalyst can, as it were, "rouse affinities which are otherwise sleeping at that temperature," by merely being present. He compares the effect of the catalyst with a rise in temperature but suggests that unlike heat, the effect of the catalyst is specific and only influences certain reactions. Finally, Berzelius discusses the role of catalytic processes in living organisms and envisages how the nourishment by way of thousands of catalytic reactions can give rise to the innumerable compounds which make up the living cell.

Of course, it can be said that already the old iatrochemists van Helmont and Sylvius had toyed with such concepts as ferment and fermentation. However, to the iatrochemists these terms had a very uncertain meaning and were far from being synonymous with modern concepts like enzyme and catalysis. For instance, much later the great German chemist Justus von Liebig (1803–1873) believed that a sugar molecule, which was being fermented to alcohol and carbon dioxide, could actually "infect" other sugar molecules in the vicinity through some kind of molecular vibrations and thus make them undergo fermentation. It was not until Berzelius defined catalysis and catalyst that terms like ferment and fermentation could eventually acquire a stringent meaning.

A Man of the World

Berzelius was very far from being an unworldly scientific recluse. On the contrary, he moved in very exalted circles including the King and his court. Chemistry was something of a fashionable science in those days and even the royal princes regularly attended the experimental work in Berzelius' laboratory. He had a number of scientific friends and pupils on the Continent of Europe and also in England. A steady flow of young scientists, especially from the German states, came to work with Berzelius and this added to his growing fame abroad. Perhaps the best known example is the famous German chemist Friedrich Wöhler (1800–1882) who was not only a pupil and collaborator, but also an intimate friend. In his autobiography he paints a vivid picture of his first encounter with Berzelius, who had impressed him not only by his gentlemanlike appearance, Berzelius made a point of always being neatly dressed, but also by his ability as a linguist. Berzelius was very fond of Wöhler; of his many pupils a considerable number came to occupy leading positions at different universities, but Wöhler was obviously the one closest to the heart of the old master. He reciprocated by spending a lot of time translating Berzelius' numerous scientific writings into German, which certainly helped to increase Berzelius' dominating position in European chemistry.

Berzelius was very much a man of the Enlightenment with its rationality and belief in human reason. His own time, the romantic era with its tearful sentimentality and speculative natural philosophy, must have been alien to him. On the other hand, he did not stand alone in his rejection of uncritical speculations and unfounded generalizations. In experimental scientific disciplines like physics and chemistry, the traditions and thoughts of the Enlightenment still lived on, and the same may be said of such basic medical sciences as physiology and biochemistry. Here, Berzelius can be seen as the custodian of a great and indispensable tradition. At the same time, he is astoundingly modern in his whole attitude with his strict rationality and his constant emphasis of the paramount importance of the experiment. To a modern scientist he is something of an older colleague, much greater and more important than we are, to be sure, but at the same time one of us.

FERMENT OR VITAL FORCE

Vitalism and Natural Philosophy

Of the medical systems originating in the 18th century, "vitalism" would turn out to be the most enduring one. It had its roots both in the ancient teachings of Hippocrates and the more recent notions about the role of the spirit, or anima, that had been proposed by Georg Ernst Stahl. An early advocate of vitalism was the French physician Théophile de Bordeu (1722–1776), who spent most of his career in Paris, but had been educated in Montpellier that was to become one of the strongholds of vitalism.

Bordeu was above all fascinated by the glands whose function he considered to depend on a mystical vital force, hence the name vitalism. This somewhat obscure philosophy was of course in marked contrast to the ideas of the iatrophysicists with their simple, mechanical explanations of body functions. To give an example, they saw the glands as a kind of sieves through which constituents of the blood could pass and give rise to different secretions. Nothing fancy or mystical about that! Against this view of the body as a sophisticated machine, Bordeu maintained that its functions were in principle dependent on vital forces present only in the living organism. Consequently, they could not be reproduced outside it. In the next century, these views would lead to a bitter conflict between the representatives of vitalism and more chemically oriented scientists.

Bordeu may seem somewhat vague and romantic in his concept of a vital force, but at the same time he had a remarkable ability to intuitively anticipate great discoveries that still lay in a distant future. For instance, he suggested that all major organs in the body function as glands by secreting substances specific for each organ to the blood, thus influencing each other by way of the circulating blood. He also proposed that it was the secretions of the gonads that were responsible for the sexual characters that developed in puberty. It would certainly seem that here Bordeu in an ingenious way foresaw what we now call endocrinology.

Paul Joseph Barthez (1734–1806) was professor of medicine at Montpellier and a prominent figure in French vitalism. He had been a prodigy in school and as a professor he made contributions not only to medicine but also to philosophy and jurisprudence. During the French Revolution, he fell out of favor for having defended the rights of the aristocracy. However, Napoleon made him, together with Jean Corvisart, physician-in-ordinary to the Emperor, which gave him a leading position in French medicine. Nevertheless, his writings about physiology and pathology, although formally brilliant, do not add much new knowledge to these subjects. His younger colleague in vitalism, Philippe Pinel (1745–1826), became famous, at least in the eyes of posterity, mainly because of his deeply humanitarian

Théophile de Bordeu.

Theophile de Bordeu (1722–1776).
Courtesy of Bonnierförlagen, Stockholm, Sweden.

Friedrich Wilhelm Schelling (1775–1854).
Courtesy of Bonnierförlagen, Stockholm, Sweden.

and courageous efforts to improve the miserable conditions and brutal treatment of the insane in French mental hospitals.

Natural Philosophy

The ideas of vitalism had been received with enthusiasm in Germany where the concept of "Lebenskraft" (vital force) was introduced by the brain anatomist Johann Christian Reil (1759–1813) and became immensely influential as a manifesto of German vitalism. From the very beginning, it had a much more metaphysical and mystical character than the French variety and gradually it became the medical branch of natural philosophy, a school that flowered in Germany at the beginning of the 19th century. Its leading mind was the philosopher Friedrich Wilhelm Schelling (1775–1854), one of Kant's most prominent pupils. Already at the age of 22, while still a student, Schelling published a book that was destined to become a canon of natural philosophy, *Ideen zu einer Philosophie der Natur.* His standing as a leading young light of philosophy is born out by the fact that he was invited to a court ball in Weimar to celebrate the turn of the century in 1800 together with such luminaries as Goethe and Schiller.

In Schelling's opinion, there was no real boundary between living organisms and inanimate material; they were all part of the same universe and all possessed of a soul. A leading principle of natural philosophy according to Schelling was the polarity that characterized the whole universe, which was perceived as having a positive and a negative pole. The male character was influenced mainly by the positive sun, while the female nature was mostly affected by the negative earth. A disturbance of this natural polarity might cause disease. On the other hand, illness could also be seen as the result of a fall from a higher to a lower level of a hierarchical creation with Man at the top of the biological ladder. These fanciful theories were never accepted in England and France, but in the German-speaking countries and in Scandinavia they came to dominate medical thinking at the beginning of the 19th century. However, not everyone was captivated by the grandiose and obscure visions of natural philosophy. The Swedish chemist and physician, Jacob Berzelius, gave the following stern verdict: "Its basis is ignorance of everything factual, love of poetry and the arts, together with a naive and ill-considered indulging in personal opinions, that by incomprehensibility have earned the reputation of being profound." Clearly, as far as Berzelius was concerned, the future of animal chemistry lay in the application of the basic principles of chemistry in order to investigate the different components that made up the living organism and in particular its ultimate building stone — the cell.

Animal Chemistry

Even a very superficial analysis of both plants and animals revealed the presence of material belonging to the categories fat and carbohydrate, to use the modern terms; substances which had played a major role in human husbandry since time immemorial. However, there was also another group of biological substances, which were recognized early on by such phenomena as the curdling of milk when acidified and the solidifying of an egg when heated. Another early observation was that blood coagulated outside the body, so that bleedings from scratches and small wounds spontaneously ceased. The formation of threads of fibrin in blood left to coagulate in a bowl must also have been previously noticed. It was phenomena like these that were used, even as late as in the beginning of the 19th century, in order to characterize a group of substances referred to as *albuminoid.*

Proteins and Amino Acids

Albuminoid substances had been shown by Scheele to contain sulfur, and in 1785 the French chemist Claude Louis Berthollet demonstrated that treatment with nitric acid made them release nitrogen. Because of the superficial likeness between albuminoid substances from animals and plants in terms of such easily observable properties as the effect of heating or acidification, it was early on assumed that substances like milk casein, egg albumin and blood fibrin occurred also in plants. Generally speaking, the number of individual proteins (to use the modern term) in nature was believed to be rather limited and it would be a long time before the extreme diversity of albuminoid substances was fully realized. Nevertheless, during the first decade of the 19th century there appeared a number of books dealing with the subject of animal chemistry, among them Antoine Fourcroy's *Système de Connaissances Chimiques* (1801) and Jacob Berzelius' *Lectures in Animal Chemistry* (1806–1808). An important reason for this widespread interest in what we today call biochemistry was the emphasis on agricultural problems that resulted from the endless wars, which plagued the Napoleonic period.

The Dutch chemist Gerardus Mulder (1802–1880) was the leading expert on albuminoid substances in the years around 1840, and he was the first to introduce the term *protein* to denote this group of substances. That name was in fact suggested to him by Berzelius in a letter 1838 in which he derived it from the Greek word *proteios* (chief) to signify that protein was the chief constituent of living organisms. Mulder felt that this came very close to his own ideas about protein, which he saw as "the most important of all the known substances of the

Justus von Liebig (1803–1873).
Courtesy of the Royal Swedish Academy of Sciences,
Stockholm, Sweden.

organic kingdom" and considered to be the origin of all other substances there, a kind of "primary compound" from which everything else emanated.

At this time, elementary analysis was virtually the only method to characterize a compound and from his analyses Mulder concluded that egg albumin, serum albumin and fibrin had the same content of carbon, hydrogen, nitrogen and oxygen, i.e. they had the same empirical formula in this respect. The difference between them was explained by the varying numbers of sulfur and phosphorus atoms that they contained in addition to the fundamental unit they had in common. He believed that herbivorous animals obtained their protein directly from the plant kingdom, while the carnivores got their protein from the same source, but indirectly by preying on the herbivores. These easy generalizations were widely accepted during the 1840s, even by such an authority as Justus von Liebig (1803–1873) who also found that proteins from animals and plants were surprisingly alike. Only the old Berzelius was critical and called Mulder's theories "an easy kind of physiological chemistry created at the writing desk." However, by mid-century the enormous diversity of proteins was beginning to be realized and Mulder's ideas about protein from the plant kingdom as the direct source of all other protein was being discarded. The only thing that lived on was the name protein, which slowly replaced "albuminoid substances" and other terms, yet it would take until the end of the 19th century before the new term was in common use.

The realization that the living organism is made up of a great number of individual proteins with different properties owed much to the discovery that they could be fractionated by taking advantage of their varying solubility in salt solutions of different concentrations. The pioneer here was the French biochemist Prosper Sylvain Denis (1799–1863) who in the 1850s published his results with salt fractionation of serum proteins. Salt fractionation became one of the most widely used methods for the separation of proteins and it remains a standard procedure to this day.

It had been noted early on that proteins could be cleaved by acid or alkaline hydrolysis, yielding what was later called amino acids. Leucine was the first amino acid isolated after protein hydrolysis (1819) followed in 1820 by the simplest of all amino acids, glycine. A number of more or less brutal methods were used for the degradation of proteins but it was soon realized that acid hydrolysis was the least damaging to the desired end-products, the amino acids. In 1846 Liebig obtained crystals of the first aromatic amino acid, tyrosine. At the end of the 19th century, a dozen amino acids had been isolated in pure form, but the list of 20 standard amino acids in protein was not completed until 1936 when threonine was discovered.

Friedrich Wöhler (1800–1882).
Courtesy of the Royal Swedish Academy of Sciences,
Stockholm, Sweden.

Emil Fischer (1852–1919).
Courtesy of the Nobel Foundation, Stockholm, Sweden.

We owe much of our understanding of amino acid and peptide chemistry to that great pioneer of organic chemistry, Emil Fischer (1852–1919). He had previously made a great name for himself in carbohydrate chemistry, but during the last two decades of his life he brought all his synthetic skill and immense knowledge of organic chemistry to bear on the building blocks of proteins, the amino acids, their structure and how they are linked to each other to form peptides. In a series of papers in 1906/07, he described the synthesis of a number of amino acids as well as polypeptides containing more than a dozen amino acid residues. Thus he proved not only the structure of the synthesized amino acids but also the nature of the peptide bond that holds them together in proteins. He considered the size of his polypeptides to approach that of natural proteins and was loathe to accept the much higher molecular weights that were now beginning to be reported for such molecules.

Early work in the 1860s and 1870s with semi-permeable membranes had led to the formulation in 1885 by Jacobus Henricus van't Hoff (1852–1911) of the equation for osmotic pressure. This property as well as freezing-point depression, was soon used for the determination of the molecular weights of dissolved substances. When these measurements were extended to include also the proteins, which had early on been designated as belonging to the class of colloids, as distinct from crystalloids such as salts and cane sugar, the results were indeed astonishing. For instance, in 1891 measurements of the freezing point depression obtained with egg albumin indicated a molecular weight of 14,000 for this molecule, and in 1905 osmotic pressure determinations gave a value of 48,000 for hemoglobin. These were shocking figures indeed, and in 1907 an incredulous Emil Fischer charged that such calculations rested on a very insecure basis since there was no guarantee for the chemical homogeneity of natural proteins, which might be mixtures of substances of much simpler composition.

Nevertheless, the exciting world of biological macromolecules was slowly being revealed and shown to include, not only proteins but also a new class of molecules, the nucleic acids, which would eventually prove to have molecular weights many orders of magnitude greater than those that had seemed so shocking to Emil Fischer.

The Meaning of Ferment

As we have noted previously, the word ferment had a very different meaning to the old alchemists as compared to, for instance, the sober and critical Boerhaave. Even in the beginning of the 19th century, ferment was still an enigma that admitted of many alternative interpretations. To give an example, in the year 1800 what remained of the French Academy of Sciences after the upheavals of the Revolution, announced a competition where the participants were asked to submit an essay in answer to the question of what distinguished the ferment from the substance being fermented. The prize for the best answer was a one-kilogram medal of pure gold. Unfortunately, the competition had to be called off because of lack of funds but the question posed is nevertheless interesting. To understand the reasoning of the Academy, one must realize that at the time the scientific community still entertained the hypothesis that a sugar molecule being fermented to alcohol and carbon dioxide could transmit this process to other sugar molecules in the vicinity by some kind of "molecular vibrations." In other words, infect them with fermentation as if it were a contagious disease.

The Advent of Microorganisms

Girolamo Fracastoro, the great poet and medical sage, had intuitively predicted the existence of microorganisms, but the first one to actually see and study them was the Dutch optician Anthonie van Leeuwenhoek (1632–1723) who used microscopes of his own construction with remarkably high resolution to examine all sorts of biological material. In 1680, he reported to the Royal Society that he had observed "little globules" in yeast with the aid of his microscope, a discovery that was ignored at the time, but which would take on great significance later on.

Lavoisier's investigations of vinous fermentation, as well as later work by the French chemist Joseph Louis Gay-Lussac (1778–1850) had demonstrated that fermented sugar was converted into carbon dioxide and alcohol, and Gay-Lussac had in fact presented a chemical equation for this process in which one molecule of glucose gave two molecules each of carbon dioxide and alcohol (as represented in modern terms). Both Lavoisier and Gay-Lussac undoubtedly thought of this as a purely chemical process that in addition required the presence of oxygen, while the ferment that activated the reaction belonged to the nebulous group of albuminoid substances. This "chemical" theory of fermentation often assumed that degradation and denaturation of these putative albuminoid substances was an essential part of their function as ferments. The influential Justus von Liebig, for instance, advocated this theory.

In 1837, the chemical explanation of fermentation was dealt a shattering blow by Charles Cagniard-Latour (1777–1859), Theodor Schwann (1810–1882) and Friedrich Kützing (1807–1893) who had independent of each other rediscovered Leeuwenhoek's old observation of the yeast cell. This remarkable coincidence was due to improvements in microscope construction represented by the achromatic compound microscope, which had recently become available and gave much better

Theodor Schwann (1810–1882).
Courtesy of Bonnierförlagen, Stockholm, Sweden.

optical resolution than previously. Cagniard-Latour was a well-known inventor and a professor at the military school in Paris. In June 1837, he presented his discovery to the French Academy of Sciences. He described the cells that he had observed in brewer's yeast as spherical particles, which could multiply by budding and he suggested that this vegetative process gave rise to the fermentation of sugar to alcohol. Furthermore, he totally rejected the idea that ferment had anything to do with albuminous material. The yeast cell was the ferment and that was all there was to it.

Theodor Schwann was a young German physician and morphologist who already had a certain claim to fame by his discovery of pepsin, a protein splitting enzyme (to use modern terminology) in the gastric juice. When he studied the proposed role of oxygen in alcohol fermentation, Schwann discovered that it was enough to heat the air that came in contact with a sterile solution of sugar in order to prevent fermentation. He therefore concluded that fermentation was caused, not by oxygen, which was not affected by heating, but by some living organism that was present in the air. This led him to examine yeast in the microscope where he observed budding cells, which he called "sugar fungi" (Saccharomyces) and interpreted as the cause of fermentation. He suggested that the fungus fermented the sugar in order to extract from it the material necessary for fungal growth, at the same time producing alcohol as a kind of waste.

Friedrich Kützing was a science teacher in what corresponded to a high school in Germany when he made his observations of budding yeast cells. When he published his discovery he made a point of renouncing any claim to priority, "since it does not matter to science who made the discovery first." For some obscure reason he also went out of his way to pick a fight with the most renowned and venerable chemist at the time, Jacob Berzelius, whose thoughts about catalysis did not appeal to Kützing. As one might have expected, Berzelius in his very influential annual report on the progress of chemistry for 1839 made some unfavorable remarks about Cagniard-Latour, Schwann and Kützing. The latter he dismissed with the words: "I pass over his philosophy regarding the organic and inorganic, which belongs to philosophical ideas that have long ceased to exert a harmful effect on the development of the sciences." Thus, already at this time the battle was joined between the fraction that would be called "the vitalists" and their antagonists "the chemists."

A Long and Bitter Conflict

It is ironic that the chieftain of the vitalists was himself one of the leading chemists of the 19th century, Louis Pasteur (1822–1895). The son of a tanner and former

Louis Pasteur (1822–1895).
*Courtesy of the Royal Swedish Academy of Sciences,
Stockholm, Sweden.*

sergeant, one of Napoleons veterans, he grew up in the little village of Dole in the South of France and got his baccalauréat from the Royal College of Besançon where he had been a rather undistinguished pupil with the character "mediocre" in chemistry. In spite of this not very promising start, in 1848 he obtained his doctorate in science in Paris and here he made his first great discovery. It concerned the optical properties of tartaric acid and the experimental approach that he had used to tackle the problem was in many ways typical of Pasteur with its mixture of ingenuity and incredible perseverance. He had found that the racemic, optically inactive form of a certain tartrate gave equal amounts of left-handed and right-handed crystals, which were mirror images of each other and could be distinguished by inspection. Pasteur now undertook the gruelling task of picking individual crystals of each sort in sufficient quantity for an analysis of their optical activity. He then found that they showed the same degree of optical activity but with opposite rotation so that the activities cancelled each other in the racemic form. This discovery gave his career a flying start. He was rewarded with the Legion of Honor, the same coveted order that his father had been decorated with for valor on the battlefield, and in 1854 he became professor of chemistry at the University of Lille.

When Pasteur turned 70 and was celebrated by scientists all over the world, the great English surgeon Lord Lister in his eulogy summed up the whole immense life's work of Pasteur by saying that medicine owed more to him than to anyone else, a judgement that can hardly be questioned. However, in what follows we will concern ourselves exclusively with his work on fermentation and his role as the leading figure among the vitalists.

Already during his stay in Lille, Pasteur had become interested in the problem of fermentation and he had early on taken the view that this phenomenon was caused by microorganisms, as had already been suggested by Latour, Schwann and Kützing. Thus, he rejected the opinion of leading chemists like Liebig and Berzelius that this was essentially a chemical reaction catalyzed (to use Berzelius's newly coined term) by ferments belonging to the albuminoid substances. Liebig had even suggested that these putative albuminoid ferments were themselves altered in the process and might, for instance, form a precipitate as an essential step in the fermentation. Pasteur's first investigations concerned fermentation to produce lactic acid, for instance in the souring of milk. He claimed that, just as in the case of alcoholic fermentation, there is what he called a "lactic yeast" always present when sugar is fermented to lactic acid. He also maintained that "lactic yeast" (or Lactobacillus in modern terminology) is a living organism and that its fermenting activity is part of its life processes.

In October 1857, Pasteur became director of scientific studies at the École Normale in Paris and here he took up the study of alcoholic fermentation. In a series of publications, culminating in a lengthy report in 1860, Pasteur established that alcoholic fermentation of sugar produced not only alcohol and carbon dioxide but also varying amounts of succinic acid and glycerine as well as trace amounts of several other substances. In view of this complexity, he rejected a purely chemical explanation of alcoholic fermentation and instead emphasized that it represented the activity of a living organism and could not be reproduced outside the yeast cell. He also showed that yeast could grow on a medium containing, besides pure cane sugar, only an ammonium salt and the minerals obtained by incineration of yeast. Under these conditions, the growing yeast fermented the sugar to alcohol and carbon dioxide. Of particular importance was, in Pasteur's opinion, the fact that no albuminoid substances of the kind required by Liebig's theory, seemed to be necessary to sustain fermentation in this simple medium.

Pasteur's studies of fermentations that yielded acidic end-products such as acetic and butyric acids, convinced him that fermentation could be described as "life without air," to use his own words. He claimed in 1861 that brewer's yeast, although it grew vigorously in the presence of oxygen, did not ferment sugar to alcohol under these conditions. On the other hand, when deprived of free oxygen the yeast fermented sugar to alcohol thereby availing itself of the bound oxygen present in the sugar molecule. Thus, in Pasteur's opinion fermentation was a method by which the yeast utilized the chemical energy of the sugar without the consumption of free oxygen.

Pasteur became increasingly convinced that real fermentation always required the presence of living microorganisms and that chemical reactions catalyzed by soluble ferments such as the splitting of protein by pepsin or carbohydrate by diastase were in reality entirely different phenomena. He now introduced his famous concept of "fermentation proprement dite" (approximately, "fermentation in the proper sense of the word"), which he defined in such a way that he excluded all fermentation processes that did not include microorganisms. Armed with this marvellous example of a circular argument he now attacked his opponents, the chemists.

The Chemical View

When the concept of "vital force" was introduced at the end of the 18th century by the German anatomist Johann Reil, it was taken to mean the sum of the mysterious forces in the living organism that enabled it to produce all those substances that it needed and which the chemists seemed incapable of synthesizing in their test tubes. The English physician John Hunter put it in a nutshell when he wrote in 1786: "No chemist on earth can make out of the earth a piece of sugar, but a vegetable can do it." On the other hand, the term vital force was rather vague, to say the least, and already in 1813 Berzelius had sarcastically pointed out that this was "a word to which we can affix no idea." Furthermore, in 1828 Friedrich Wöhler was able to prepare urea from ammonium cyanate, thereby giving the lie to Hunter's claim that the chemists could not synthesize material normally produced by a living organism. However, a champion of the vitalists, the German physiologist Johann Müller, nimbly avoided what ought to have been a crushing blow by pointing out that urea could scarcely be "considered as organic matter, being rather an excretion than a component of the animal body."

The constantly increasing number of soluble ferments, on the other hand, represented a serious threat to the central dogma of vitalism. Pasteur and other protagonists of vital force were prepared to accept the idea that cells, for instance microorganisms, contained "organized ferments" that explained their ability to ferment sugar to alcohol, etc. However, the soluble (or "unorganized") ferments were something entirely different and more problematic. We have already encountered pepsin and diastase, and in 1837 the two life-long friends Liebig and Wöhler discovered an albuminoid substance in almonds that was able to hydrolyze amygdalin from bitter almonds into benzaldehyde, hydrogen cyanide and glucose. They called this soluble ferment emulsin and considered it particularly interesting because its substrate, amygdalin, had been obtained in pure form and its structure was later determined.

It must be emphasized that Pasteur and his vitalism was seriously criticized not only by German scientists like Liebig and Wöhler but also by the well-known French chemist Marcelin Berthelot (1827–1907), who in 1860 had obtained a soluble ferment (invertase) from yeast that catalyzed the splitting of sucrose to a mixture of glucose and fructose (invert sugar), hence the name of the enzyme. Pasteur lost no time in pointing out that this reaction was brought about by a great many other substances besides invertase, for instance by all the acids. Clearly, in his opinion the splitting of cane sugar was not worthy of being called fermentation. Berthelot, on the other hand, had categorically rejected "life" as a meaningful explanation of such phenomena as fermentation. In his view, one should instead use chemical methods and look for what we would call molecular explanations of biological processes. Berthelot's relations with Pasteur were not improved when in 1878, after the death of the great French physiologist Claude Bernard, he arranged for the posthumous publication of a manuscript in which Bernard claimed to have demonstrated what seemed to be alcoholic fermentation in a cell-free preparation from the juice of rotting fruit. Because of the immense authority of Bernard this manuscript, which was probably not intended for publication in its present preliminary form, placed Pasteur in a rather awkward position. In fact, he had to spend much time and effort in order to show that Bernard had made a mistake here.

Marcellin Berthelot (1827–1907).
*Courtesy of the Royal Swedish Academy of Sciences,
Stockholm, Sweden.*

In 1870, Pasteur's old adversary Justus von Liebig again challenged what some authors called the "French vitalist theory" at the same time that the old conflict between Prussia and France flared up into a full-scale war that would turn out to be a catastrophe for France. Pasteur was a great French patriot and he showed no mercy in his attacks on the old and ailing Liebig, who died in 1873. Perhaps Liebig's life was not actually shortened by the fight with Pasteur, but it certainly did not brighten his last years. At the time Pasteur seemed to be the victor, although his view of fermentation as "life without oxygen" had to be abandoned. As we shall see in the following the soluble ferments would eventually prevail even in the case of alcoholic fermentation.

The Lock and Key

At the time of Pasteur's quarrel with Berthelot over the publication of Bernard's posthumous manuscript, the number of the controversial soluble ferments approached 20, and in 1878 the German biochemist Willy Kühne (1837–1900) suggested that such ferments should be called enzymes. At the same time, he carefully avoided taking sides in the conflict between vitalists and chemists, the new term was just a question of convenience. The name enzyme was not an immediate success. The leading German biochemist Felix Hoppe-Seyler (1825–1895), for instance, said scornfully: "The new word enzyme could be added to the large number of new names that Kühne has proposed for totally unknown substances." Nevertheless, with time the word enzyme would completely replace ferment and is now the only term used.

We have already encountered Emil Fischer in his role as the leading figure in organic chemistry at the end of the 19th century. However, he was also a pioneer in our understanding of the mechanism of enzyme action. A number of different hypotheses had been advanced previously in order to explain this problem. When in 1835 Berzelius defined the term catalysis he envisioned the catalyst as being active simply by its presence, without actually participating directly in the reaction or interacting with the reactants. His friend and protégé, the German physical chemist Eilhard Mitscherlich (1794–1863) thought more in terms of contact between the ferment and what we now call its substrate, as a prerequisite for fermentation. On the other hand, according to Liebig the ferment not only participated in the reaction, it was itself actually changed in the process so that it appeared afterwards as a precipitate. He also subscribed to the old idea of the reaction being propagated by molecular vibrations in the substrate molecules.

In a famous paper with the title *Bedeutung der Stereochemie für die Physiologie* ("The Importance of Stereochemistry for Physiology"), Fischer put

Eilhard Mitscherlich (1794–1863).
*Courtesy of the Royal Swedish Academy of Sciences,
Stockholm, Sweden.*

forward in 1898 the first full-fledged model for the mechanism of enzyme action, his celebrated lock-and-key hypothesis. At the time, nothing was known about the three-dimensional structure of enzymes and their active sites, where the substrates are bound as part of the catalytic process. Fischer's idea of a lock-and-key mechanism was therefore intuitive, or rather an educated guess based on his previous experience regarding the role of stereochemistry in molecular interactions in organic chemistry. In any case, both studies of enzyme kinetics as well as the extensive knowledge of enzyme conformation that we now have acquired through X-ray diffraction analysis, have confirmed the main features of Fischer's model. The substrate is indeed structurally adapted to a particular site on the surface of the enzyme in much the same way as a key fits into a lock. From this basic concept follows the assumption that the formation of an enzyme-substrate complex is an obligatory step in the catalytic process. Furthermore, the results of enzyme kinetics are consistent with this step being the rate limiting in catalysis.

A Fortuitous Observation

The Man from Upper Bavaria

Eduard Buchner (1860–1917) is in many ways the most unlikely of the great biochemical pioneers. He came from a family of scholars but there was something rustic and unsophisticated about his whole appearance and he never lost his characteristic Upper Bavarian accent. His father was a professor of forensic medicine in Munich and Eduard and his elder brother Hans were children of his third marriage. Eduard's father died when the boy was 12 years old and it would appear that this tragic loss also drastically weakened the economy of the family. After graduating from the Gymnasium in Munich, instead of going on to the university as the family traditions dictated, Eduard joined the 3:d Bavarian field artillery regiment with the intention of becoming an officer. However, although he obviously liked the life of a soldier, he left the army after having served for a year, perhaps because of his weak economy. Instead he took up a position in a cannery factory in Munich where he worked in a modest capacity for over four years until he was rescued by the intervention of his brother. Hans was ten years older than Eduard and would seem to have taken the place of the dead father in the life of his younger brother. He had chosen a medical career where he would distinguish himself as a hygienist and eventually become the successor of the well-known Munich professor Max von Pettenkofer.

Encouraged and probably also financed by his brother, Eduard entered the University of Munich where he studied chemistry under the famous Adolf von Baeyer and plant physiology in the laboratory of Karl von Nägeli. This resulted in his first publication on alcoholic fermentation in 1886 in which he concluded that, contrary to Pasteur's views, oxygen was not an inhibitor of fermentation. In the department of chemistry, he made the acquaintance of the organic chemist Theodor Curtius and they became close friends and collaborators. In 1888 he obtained his doctorate with Baeyer, although in reality Curtius would seem to have been his main supervisor. Two years later, he was appointed teaching assistant and the next year followed his "Habilitation" which made him eligible as "Privatdozent," an unsalaried position but at the same time a necessary step in his academic career. His research was at this time almost exclusively concerned with classic organic chemistry, and one gets the impression that Adolf von Baeyer was somewhat doubtful about Buchner's future as a scientist.

In 1893 Buchner accompanied his friend Curtius to Kiel and in 1896 another friend, Freiherr von Pechmann, saw to it that Buchner was appointed professor of analytical pharmaceutical chemistry at the University of Tübingen. It was here that he published his first pioneering work on cell-free alcoholic fermentation in 1897. We will consider his work on fermentation in more detail in a following section. His contribution to this field was generally recognized as being important

Eduard Buchner (1860–1917).
Courtesy of the Nobel Foundation, Stockholm, Sweden.

Friedrich Althoff.
Source: Author's collection.

and he was called to the Landwirtschaftliche Hochschule (College of Agriculture) in Berlin as professor of general chemistry in 1898. At the same time, he became director of Das Institut für Gärungsgewerbe (The Institute for Fermentation Industry).

During his stay in Tübingen he had made the acquaintance of his future wife, Lotte Stahl, the daughter of a professor of mathematics and the granddaughter of the famous philosopher Friedrich Trendelenburg. They were married in August 1900 and Buchner's friend and future biographer, Carl Harries, gave an amusing account of the wedding. The bridegroom was obviously in excellent spirits and during the festivities he apparently became slightly intoxicated. In any case, while on his way home after the traditional drinking bout the night before the wedding, he struck up a song on the peaceful streets of Tübingen with a voice so penetrating that he came very close to being arrested by the police. One wonders what the distinguished academic family that he was marrying into, would have thought if the bridegroom had to be fetched directly from the police station to the wedding ceremony.

From a scientific point of view, Buchner's time in Berlin was very productive. He nevertheless, seemed to have longed for a professorship at one of the leading German universities. During this period, the question of who got to become a professor was in reality managed by a very influential high official in the Prussian Kultusministerium (Department of Education) by the name of Friedrich Althoff. Because of his work on fermentation, Buchner's name was often put forward when professorships became vacant. In 1904, this happened again and it resulted in a meeting between Buchner and Althoff which turned out to be a catastrophe. The encounter certainly shows that whatever social graces Buchner may have possessed did not include a talent for diplomacy in the handling of influential superiors.

His Excellency (Althoff had this exalted title in the Prussian hierarchy) had demanded a meeting with Buchner to discuss a chair in chemistry that had recently become available in Königsberg. Buchner, being pressed for time on that particular day, somewhat haughtily declared that he could only come if Althoff received him immediately so that he did not have to wait for the audience. This Althoff promised to do; he would not let him wait for more than 10 minutes. However, when Buchner arrived at the Kultusministerium he was kept waiting for over an hour and when Althoff finally showed up for the meeting, Buchner who was a stickler for punctuality, was highly irritated. He was sitting there, clock in hand, and demonstratively pointing to the dial he asked: "Is that ten minutes?" This was not the way a lowly professor should address such a lofty person as Althoff and his Excellency irately replied: "Obviously, you do not want to go to Könighberg!" and stormed out of the room.

As might have been expected, there were no further openings for Buchner while Althoff was in charge at the Kultusministerium and he remained at the Agricultural College until 1909 when Althoff's successor at the Ministry saw to it that Buchner was appointed to a chair in chemistry at the University of Breslau. On the other hand, it was during this period of frustrated academic ambitions that Buchner received the Nobel Prize in Chemistry in 1907 for his work on cell-free fermentation (the Nobel Prize will be treated in more detail below). After two years in Breslau he had the satisfaction of being called to a chair in Würzburg, which meant that he returned to his beloved Bavaria. This was a great joy to Buchner, who was not only a local patriot but also a German nationalist with somewhat chauvinistic inclinations. In recognition of his Nobel Prize, the students at the Agricultural College celebrated with a torchlight procession where he gave an address in which he talked contemptuously about the strive for peace and brotherhood between nations. Instead, he spoke glowingly about patriotism and praised war as a purifying thunderstorm, which promoted the utmost manly prowess and devotion to Kaiser and fatherland. This was heady stuff even at the time when the nations, which would soon be involved in the Great War, seemed completely oblivious of the disaster they were heading for.

In August 1914, the threatening storm finally broke. The British Foreign Secretary, Sir Edward Grey, looking out over London as the street lamps were being lit, made his prophetic remark: "The lamps are going out all over Europe; we shall not see them lit again in our lifetime." Eduard Buchner was not haunted by any such dark forebodings when he joyfully, at the age of 54, reported for military duty in the Bavarian artillery. He was put in charge of an ammunition supply unit with the rank of captain and participated in campaigns both in Northern France and on the eastern front. Buchner was decorated with the Iron Cross, second class, and in 1916 he was promoted to major. However, his faculty in Würzburg wanted him back in his civilian capacity as professor and they seemed to have prevailed on the military authorities to discharge him. A year of civilian life proved to be enough for Buchner, he longed to be back at the front again, fighting for Kaiser and fatherland. To quote his biographer Harries, in June 1917, "happy as a youngster he went to the front in Rumania," where on 11 August he was wounded in the left thigh by shrapnel. He was hospitalized in Focsani but the wound was not considered very serious to begin with, and the next day he wrote optimistically to his wife that everything had gone very well. However, in the night of August 13 the medical orderly on duty heard a deep sigh from Buchner's bed and found that he had suddenly passed away.

His friend, Carl Harries, wrote an obituary of him but the catastrophes of the Great War of course tended to overshadow everything else, and the death of a German professor was easily overlooked at the time. Buchner could with some justification be called a simple soul and as we shall see in the following, some of his famous colleagues did not have a very high opinion of him as a scientist. Yet, he had made one of the most important observations in the history of biochemistry and had, on the whole, drawn the correct conclusions from it.

The Fermenting Extract

Eduard Buchner's elder brother Hans was a rather impressive figure. Carl Harries, who was one of the guests at Eduard's wedding where he met Hans, described him as a very tall and fine-looking man with grave but friendly eyes, eminently eloquent and in every way very different from Eduard. Apparently, compared to his elder brother, Eduard seemed much more unassuming, not to say insignificant. It is easy to understand that he looked up to his brother and that Hans had a considerable influence over him.

In the 1890s, Emil von Behring and his collaborators had discovered the ability of animals infected with diphteria and tetanus bacteria to form antibodies against the corresponding bacterial toxins and such antitoxins were beginning to be used therapeutically with increasing success. At the same time, Robert Koch erroneously reported that tuberculin, a substance isolated from cultures of the tubercle bacillus, could cure early cases of tuberculosis. This led Hans Buchner to believe that perhaps extracts of harmless microorganisms could be used to produce medically active substances. He thought of yeast as an easily available microorganism that might be useful, but unfortunately yeast cells proved to be very tough and hard to disintegrate. He had managed to interest Eduard in the project and to begin with it was a question of finding an effective and not too laborious method to produce a yeast extract. At the suggestion of Martin Hahn, an assistant of Hans Buchner, they ground yeast in a mortar with one part of pulverized quartz and one-fifth of diatomaceous earth (kieselguhr) so as to obtain a paste that was then rapped in canvas and subjected to a pressure of 90 kilograms per square centimeter. In this way, one kilogram of yeast yielded approximately 500 milliliters of fluid.

There was a problem with the preservation of the fluid, which seemed to decompose very easily on storage. Hans then thought of the classical method of preserving fruits by adding a high concentration of sucrose. In 1896 Eduard, having joined his brother during the vacation from Tübingen, was present when the press-juice was preserved with sucrose. He then made the momentous observation that after the adding of the sucrose there was a lively formation of gas in the fluid. In all probability, his brother must have noticed a similar phenomenon in earlier experiments of the same kind, but apparently had paid no attention to it. However, it would seem that Eduard immediately drew the correct conclusion. Their yeast extract had the ability to ferment sucrose to alcohol and carbon dioxide.

There had been numerous unsuccessful attempts in the past to prepare yeast extracts containing soluble ferments that could produce alcohol from sugar. For instance, in 1846 Friedrich Wilhelm Lüdersdorff had ground a small amount of yeast, which he had placed on a glass plate, until no intact yeast cells were visible in the microscope. He had not been able to demonstrate any alcoholic fermentation in the resulting paste and Carl Schmidt reported the same result in similar experiments. On the other hand, in 1872 and later in 1897 Marie von Manassein claimed that she had observed cell-free alcoholic fermentation in a yeast extract obtained by extensive grinding with fine sand. However, her experimental methods were seriously criticized, particularly with regard to the possible presence of intact yeast cells in the extract and her results were not generally accepted. On the whole, we must conclude that Eduard Buchner's publication in 1897, *Alkoholische Gährung ohne Hefezellen* ("Alcoholic Fermentation Without Yeast Cells") represents the first reliable report of cell-free alcoholic fermentation.

Buchner called his yeast extract "zymase" and his claim that it would catalyze fermentation of sugar to alcohol met with surprisingly little opposition from the vitalists that had for so long dominated this hotly contested question. Perhaps it had something to do with the fact that the great chieftain of vitalism, Louis Pasteur, had died only two years earlier in 1895. On the other hand, maybe the notion of soluble ferments (or enzymes as they were now increasingly being called) and their paramount importance in the metabolism of the cell, had simply reached a stage of maturity where it had become generally acceptable to the scientific world. It had taken the better part of the 19th century for this idea to become respectable but now the time was obviously ripe. This does not mean that there were no objections at all. On the contrary, the greater efficiency of the intact yeast cells in alcoholic fermentation compared to a corresponding amount of zymase was repeatedly noted and the elusive concept of "living proteins" was introduced as an argument against the soluble ferments. In fact, it was suggested that Buchner's zymase was active only because it contained fragments of protoplasm. In a series of publications Buchner was able to dismiss these objections, but one of Adolf von Baeyer's most outstanding pupils, the future Nobel laureate Richard Willstätter, would nevertheless seem to have subscribed to such views as late as in the 1930s. In any case, he was obviously not an admirer of Buchner's, and in his autobiography Willstätter cites his famous teacher von Baeyer as having said, when he learned of Buchner's discovery: "This will make him famous, even though he has no talent as a chemist."

Nevertheless, in spite of Adolf von Baeyer's rather unflattering remarks about his former pupil, there can be no doubt that Buchner's somewhat fortuitous

Adolf von Baeyer (1835–1917).
Courtesy of the Nobel Foundation, Stockholm, Sweden.

Richard Willstätter (1872–1942).
Courtesy of the Nobel Foundation, Stockholm, Sweden.

discovery came at exactly the right moment and forever laid the disturbing ghost of vitalism to rest. Never again would we have to consider mystical powers that sustain unique processes, which can only take place in the living cell and cannot be reproduced in the test tubes of the biochemists. The history of cell-free fermentation teaches us that a problem is not necessarily unsolvable just because it has not yet found its solution, to paraphrase Jacque Loeb.

THE NOBEL PRIZE

Idealism and Dynamite

Nowadays, Alfred Nobel is something of a celebrity and every year, when his prizes are awarded to scientists, authors and peace workers on the 10th of December, the public is reminded of this shy and intensely private man, who would surely have been made very uncomfortable and embarrassed by all the speeches of appreciation and homage given in his honor, had he been able to listen to them. The fact is that during his lifetime he was almost unknown to the general public and there is every reason to believe that this was the way he wanted it. Who was he then and what made him create the celebrated prizes that have made his name famous?

Alfred Bernhard Nobel was born in Stockholm on 21 October 1833. At the time, it was believed that the family was of foreign origin and Nobel's own father, the inventor and industrialist Immanuel Nobel, was of the opinion that he was descended from an English clergyman who had moved to Sweden and settled there. This family tradition is, however, a complete fabrication and has no substance whatsoever. Instead the Nobel family came from farmers in Scania, the southernmost part of Sweden, and the name was originally Nobelius, derived by Latinization of the name of the parish, Nöbbelöv, where they resided. In the 17th century it was customary among many young Swedes, who had received a university education, to Latinize their names and this Petrus Nobelius had done. He also married the daughter of Olof Rudbeck, anatomist and generally speaking the leading mind and all-round genius of the University of Uppsala. It was from this union Alfred Nobel was descended.

His father, Immanuel Nobel, had little enough schooling, knew no foreign languages and could barely write. Nevertheless, he was intelligent and ingenious, full of ideas and sometimes rather fantastic projects. He was apprenticed to a builder and in his spare time he attended a trade school, the only formal education he ever had. Unfortunately, when he tried to establish himself as an architect and builder, his ambitious projects were unsuccessful and in 1833, the same year as his son Alfred was born, he declared bankruptcy. In 1837, he moved to Russia and started a successful machine shop in St. Petersburg. The shop expanded and became very profitable, particularly after the outbreak of the Crimean war when he was awarded a number of Government contracts. However, when the war ended in 1856 these contracts were terminated and in 1859 he once more went bankrupt. In 1842, his wife and three sons had joined him in St. Petersburg. During the days of prosperity the sons had private tutors and Alfred, in fact, never went to school or attended a university. Like his father he could be said to have been self-educated, but he managed to become thoroughly trained as a chemist and he was

Alfred Nobel (1833–1895).
Courtesy of the Nobel Foundation, Stockholm, Sweden.

an accomplished linguist, who mastered Russian, English, German and French, in addition to his native Swedish.

When Immanuel Nobel went bankrupt for the second time, he and his wife and their younger sons, Alfred and Emil, returned to Sweden. There Immanuel again ventured into business, helped by his son Alfred. His two elder sons, Robert and Ludvig, remained in Russia where they became very successful industrialists, setting up a great petroleum industry in Baku, which made them very wealthy. In the factory that Immanuel had started after his return to Sweden, his son Alfred began to make experiments with nitroglycerine, a very dangerous explosive that the Italian chemist Ascanio Sobrero had synthesized in 1846. In 1863, Alfred was granted a patent on a percussion detonator and with that his career as an inventor can be said to have started. However, in September the next year a terrible explosion destroyed the Nobel factory with the loss of several lives, among them that of Alfred's youngest brother Emil. This last misfortune and the death of his son proved too much for Immanuel and he had a stroke from which he never really recovered.

While his father was prostrate with grief over the death of his youngest son, Alfred seemed to have kept up his energy and initiative. He organized new factories and companies for the manufacture of nitroglycerine and its transformation into a safer and more manageable explosive. In 1867, he was granted a patent on dynamite in which the nitroglycerine was mixed with a porous absorbent (kieselguhr) so as to make it safe to handle. The invention of the dynamite and his skilful management of its commercial exploitation made him an exceedingly rich man. At the same time, he seemed to have derived very little satisfaction from his wealth. It was important to him, but only as a tool for achieving certain goals as an industrialist and businessman. By nature, he appeared to have been a melancholic, a dreamer and a recluse, who often gave vent to a pessimistic, not to say misanthropic view of his fellow beings. He had few social graces and spent most of his time working in his laboratory, completely absorbed by the project at hand. He never married and we know very little about his relations to women, except that they seem to have been relatively short-lived and on the whole unsatisfactory. This is in contrast to the intellectual liaisons that he formed with, in particular, the Baroness Bertha von Suttner, who had a profound influence on his views of peace and international relations.

His feeling of loneliness was further intensified by the notion that he was a man without a homeland. He had been born in Sweden but between the age of nine and 16 he had lived in Russia. As an adult he had become increasingly a cosmopolitan with his headquarters in Paris where in 1875 he had acquired a house on Avenue Malakoff. Towards the end of his life he bought a villa in San Remo, Italy, and he also purchased a mansion in rural Sweden close to the Bofors factory where he was a major stockholder. Presumably he intended to spend his remaining years at his Swedish mansion. Thus, he can be said to have had many residences, although at the same time he felt nowhere at home. It may have been both his longing for a real home and the hope for a congenial female companion that in 1876 made him place an advertisement in a Vienna newspaper in which it said: "A wealthy, highly educated elderly gentleman, living in Paris" /looked for/ "a lady of mature years with a knowledge of languages to act as his secretary and housekeeper." The "elderly gentleman" was in fact 43 years old at the time and the "lady of mature years," who answered his advertisement, Countess Bertha Kinsky von Chinic und Tettau, was ten years younger. She came from an impoverished aristocratic Austrian family and while being employed as a governess in the wealthy von Suttner family she and the brother of her pupils, Arthur von Suttner, had fallen in love with each other. He was seven years younger than Bertha, but they had become secretly engaged and had every intention of marrying. When the parents found out about the love affair they forbade the marriage both because of the difference in age between the prospective bridegroom and his intended bride and because of the poverty of Bertha Kinsky's family. Obviously, she could not continue as governess for Arthur's young sisters and she now found herself without employment. Such was the romantic background of her application for the position as secretary and housekeeper to Alfred Nobel.

They met in Paris and it would seem that Bertha made a deep impression on her future employer. She was considered a radiant beauty at the time with an extensive education, she was musically and literary gifted with beautiful manners, not to mention that she was an accomplished linguist, who spoke French, English and Italian, besides her native German. Nobel engaged her on the spot but unfortunately, having known her for only a week, an urgent business trip made it necessary for him to leave Paris. While Nobel was away she was bombarded with despairing letters from her young lover, Arthur von Suttner, imploring her to return to Vienna. She allowed herself to be persuaded, sold a valuable diamond trinket to pay for the trip back and leaving a letter of explanation for Nobel, she set out for Vienna and her future husband. One can easily imagine Nobel's frustration when he returned to Paris, but he, nevertheless, remained in close contact with her and they corresponded frequently, particularly during his last years. They only met again twice, the last time in 1892, but there can be no doubt that Bertha von Suttner, who was deeply involved in the work for world peace, was a strong influence when Nobel decided to include a Peace Prize in his famous will.

While his personal life can hardly be described as a great success, his wealth continued to grow. Like King Midas everything that he touched seemed to turn into gold, even if there were also some economic setbacks. This was particularly true in France, where some of his associates in the French dynamite companies proved to be incompetent, not to say downright fraudulent. In the beginning of the 1890s, Nobel on certain occasions believed himself to be on the verge of bankruptcy, but through his financial ingenuity and boundless energy the threatening disaster was avoided. However, conflicts with the French authorities forced him to abandon Paris as his European headquarters and move to his Italian villa in San Remo. Another source of concern was a prolonged lawsuit in England over the patent rights to a smokeless, nitroglycerine gunpowder. Eventually, in 1894 Nobel lost his suit, something that he with some justification considered to be a clear miscarriage of justice. In fact, to give vent to his feelings he wrote a play, "The Patent Bacillus," where he satirized the British legal system and at the same time displayed a certain talent as a dramatic writer.

A Momentous Will

Alfred Nobel's health had always been delicate. In his later years, he developed a heart condition and he often complained about his poor health in letters to his relatives. Nevertheless, when the situation so required, for instance during the crisis involving his French dynamite companies, he could display an amazing energy and drive, perhaps even ruthlessness. However, in 1895 he must have had a premonition of his life drawing to an end, and on 27 November he made his will in which he provided for the prizes that have been named after him. What made him select the particular fields of human endeavor where prizes should be awarded?

Physics and chemistry are branches of scientific research that would naturally appeal to him as fields for rewards, seeing that Nobel himself was a chemist and an inventor. Physiology or medicine, as this prize was eventually called, is perhaps not quite so obvious as a field of interest for Nobel. However, in October 1890 Nobel had invited a young Swedish physiologist, Johan Johansson, to visit him in Paris in order to discuss problems in physiology and medicine that were of interest to Nobel. Johansson stayed in Paris for five months and even carried out a number of experiments in Nobel's laboratory. He was offered a position at a medical research institute that Nobel proposed to create in Paris, but Johansson could not be tempted away from his career at the Karolinska Institutet. Even so, he managed to establish a close contact with Alfred Nobel and one cannot doubt that this was important for Nobel's decision to include physiology or medicine among the fields where his prizes should be awarded. At the end of the 19th century, physiology was considered the dominating science among basic medical sciences, as illustrated by the fact that biochemistry was often called physiological chemistry. It is therefore reasonable to believe that when Nobel used the term "physiology," what he had in mind was probably all the basic sciences of medicine. In his will, the word "medicine" would then denote the clinical disciplines of medicine.

That brings us to literature; what was Nobel's interest here? In his youth he had written poetry, not in his native Swedish but in English. He was obviously strongly influenced by Shelley, whose idealism greatly appealed to him. His own poems have been judged to show considerable talent and poetic instinct. Later in life, he wrote the plays "Nemesis" and "The Patent Bacillus" as well as some novels. He was an accomplished linguist and was able to maintain familiarity with the literature on all major European languages without having to depend on translations. In spite of being constantly pressed for time by his extensive business interests and his work as an inventor, there can be no doubt that Nobel managed to be widely read and that literature meant a great deal to him. It is therefore only natural that in his will he should make provisions not only for scientific prizes but also for a literary one. However, why should an inventor of dynamite and smokeless gunpowder be interested in world peace? It is worth recalling that at the time when Nobel wrote his will it was widely believed, and not only in Kaiser Wilhelm's Germany, that wars were unavoidable and perhaps might even be seen as a strengthening and purifying experience for a nation. Obviously, Nobel did not share these martial views, but his recipe for achieving eternal peace was a somewhat original one. He refused to take part in peace congresses and in a letter to Baroness Bertha von Suttner, he turned down an invitation to attend such a meeting with the following words: "My factories may make an end of war sooner than your congresses. The day when two army corps can annihilate each other in one second, all civilized nations, it is to be hoped, will recoil from war and discharge their troops." However, in a letter of 1893 to the Baroness he had modified his views. He was now prepared to create a prize to be awarded for work that tries to induce all nations to loyally pledge themselves to turn against the initial aggressor who attempts to start a war. This would make war impossible and instead conflicts would have to be solved by negotiations. As we shall see below, he went one step further in his will and actually supported the holding of peace congresses.

Nobel's will was drawn up in his own hand without any legal help and was dated 27 November 1895, in Paris. The original is in Swedish but contains a number of minor linguistic mistakes testifying to Nobel's limited experience with the Swedish language. An English translation of the document follows here.

"I, the undersigned, Alfred Bernhard Nobel hereby declares as my last will and testament that what property I may possess at my death shall be dealt with in the following way:

The capital from my realizable estate shall be invested by my executors in safe securities and shall constitute a fund, the interest on which shall be annually distributed in the form of prizes to those who, during the preceding year, shall have conferred the greatest benefit on mankind. The said interest shall be divided into five equal parts, which shall be apportioned as follows. One part to the person who shall have made the most important discovery or invention within the field of physics; one part to the person who shall have made the most important chemical discovery or improvement; one part to the person who shall have made the most important discovery within the domain of physiology or medicine; one part to the person who shall have produced in the field of literature the most outstanding work of an idealistic tendency; and one part to the person who shall have done the most or the best work for fraternity among nations, for the abolition or reduction of standing armies and for the holding and promotion of peace congresses.

The prizes for physics and chemistry shall be awarded by the Swedish Academy of Sciences; that for physiological or medical work by the Karolinska Institutet in Stockholm; that for literature by the Academy in Stockholm; and that for champions of peace by a committee of five persons to be elected by the Norwegian Storting. It is my express wish that in the awarding of the prizes no consideration whatsoever shall be given to the nationality of the candidates, so that the most worthy shall receive the prize, whether he be a Scandinavian or not.

This will cancels all previous testamentary provisions, in case any such should be found after my death.

Finally, I direct that after my death the arteries shall be opened and when definitive signs of death have been ascertained by competent physicians, the body shall be cremated."

Creating the Nobel Foundation

During the last three years of his life, Alfred Nobel had employed a young Swedish chemist, Ragnar Sohlman, as a kind of private secretary and it is obvious that Nobel increasingly came to rely on the young man as an honest and trustworthy personal assistant, now that his own health was rapidly declining with frequent attacks of angina pectoris. In his last letter to Ragnar Sohlman, who was then in Sweden looking after his employer's interests there, Nobel says that he can only "write these lines with difficulty." It would seem that after having finished the letter he suffered the stroke that three days later, on 10 December 1896, ended his life. His servants immediately telegraphed the news of Nobel's death to Sohlman,

Ragnar Sohlman (1870–1948).
Courtesy of the Nobel Foundation, Stockholm, Sweden.

who on his return to San Remo, where Nobel had passed away, was reached by the information that he was one of the two executors of Nobel's will. It would seem that this caused him a sleepless night; maybe he had a presentiment of all the difficulties that the provisions of that will would entail. When the complete text of the will, which had been deposited in a bank in Stockholm, was sent to him a few days later, he fully realized what he was in for. However, it would seem that during the years of endless troubles and difficulties that now followed, he never wavered in his resolve to fulfil the wishes expressed in the will of Alfred Nobel, a man that he must have considered a close friend in spite of their great difference in age.

The reaction of Nobel's relatives to the stipulations in his will differed a great deal. His eldest brother Ludvig had died already in 1888 of a heart condition and Ludvig's eldest son, Emmanuel, had taken over the management of the Nobel Companies in Baku. Emmanuel was naturally very interested in the stock, amounting to a controlling interest, that his uncle had in the Nobel Brothers Naphtha Company and how that should be handled by the executors of his will. At the same time, he was very reluctant to oppose the wishes of his uncle's will. In fact, Emmanuel turned out to be the best supporter Sohlman could ask for in his future conflicts with certain members of the Nobel family. He once said something to Sohlman that made a deep impression on him: "You must always remember the obligation implied in the Russian word for the executor of a will, *Dushe Prikashshik*, which means the spokesman for the soul." Once the terms, under which the estate of Alfred Nobel should sell his petroleum stock to Emmanuel and the other children of Ludvig Nobel, were agreed upon, Sohlman had the very best relations with the Russian part of the Nobel family. However, with some of the Swedish relatives the relations of the executors were far from cordial.

The reactions in some of the Swedish newspapers on the provisions in Nobel's will, had been rather negative. In some cases, the will had even been attacked as being downright unpatriotic, since no consideration was given to the nationality of the prize candidates. Some of the papers went so far as to urge Nobel's Swedish relations to contest the will, something they were probably more than willing to do. Obviously, there were a number of issues that had to be resolved before the great project could be realized. For instance, there was the thorny but very important problem of Nobel's legal residence at the time of his death. This was significant both for the death duties and the question of where the will should be submitted for probate. Assuming that the French authorities should decide that he had been resident in Paris, they might impound his very considerable estate in France and prohibit it from being transferred to Sweden. The death duties could become enormous and the whole idea of creating a prize awarding fund might be jeopardized. On the whole, it would seem best if Nobel was found to have been resident on his Swedish mansion Björkeborn and if as much as possible of his estate, particularly that in France, could be transferred to Sweden.

All the complications in connection with Nobel's will made it necessary for Sohlman to make frequent visits to Paris where he and his wife established themselves in an old-fashioned family hotel that became his headquarters. He made contact with the Swedish Consul General, Gustaf Nordling, and secured his help in collecting all the papers relevant to Nobel's French estates and making them ready for transportation to Sweden. His consultations with French lawyers, recommended to him by Nordling, had strengthened his conviction that there was danger in delay and that every effort should be made to get Nobel's assets to Sweden as soon as possible. He became even more alarmed when he found out that two of Nobel's nephews, Hjalmar and Ludvig, the sons of Nobel's elder brother Robert, accompanied by the husband of their sister Ingeborg, Count Carl Gustaf Ridderstolpe, had arrived in Paris and were taking legal advice with the obvious intention of contesting Nobel's will. However, Sohlman had made his preparations and he was now ready to act. The securities, in postal packages and insured by the Rothschild banking firm, were collected from the vault of the bank by Sohlman and Nordling assisted by a clerk from the Consulate, and taken in an ordinary cab to the *Expedition de Finances* at the Gare du Nord and from there they were sent on to Stockholm. Several such trips were needed and since hold-ups and robberies were not unknown in Paris at this time, Ragnar Sohlman sat in the cab with a loaded revolver in his hand ready to defend Nobel's fortune against any attack.

In the spring of 1897, the rescue of the Nobel estate to Sweden having been successfully completed, the executors of his will sent identically worded letters to all the Swedish institutions, that Nobel had selected as custodians of the Nobel Prizes, in which they invited them to consultations about how the prizes should be awarded. They also sent a respectful request to the Norwegian Storting (Parliament) asking it to accept the responsibility for the Peace Prize. Already on 26 April 1897, the Storting adopted a formal resolution taking upon itself the task as set out in Nobel's will. Thus the Norwegian Parliament made no difficulties, and gladly accepted its new duties. In contrast to the readiness of the Norwegian authorities, the Swedish institutions that Nobel had wanted to entrust with the awarding of the scientific and literary prizes, raised many more objections. For the executors of the will, it was very urgent to reach an agreement with the intended prize-awarding institutions, particularly since the will had already been submitted for probate at the County Court of Karlskoga, to which Nobel's Swedish residence belonged. The court could not possibly reach a decision until the question had been settled of how and by what institutions the prizes should be awarded.

Nobel had stated in his will, that the prize in literature should be awarded by what he called "the Academy in Stockholm," which obviously meant the Swedish Academy, established in 1786 by King Gustavus III as an Academy of Letters. When the Academy became aware of its proposed role as the awarder of a very considerable literary prize, there was much hesitation among its 18 members about whether to accept this difficult responsibility. The main argument against undertaking the task was that it would involve so much work that other duties of the Academy would have to be neglected. However, the Permanent Secretary of the Academy, Carl David af Wirsén, a poet of a rather conservative literary taste, took a very different view. He strongly supported Nobel's idea and urged the Academy to take on its new role as an arbitrator of the prize in literature. Being a man of considerable influence, he managed to overcome the opposition and by a final vote of 12 to two the Academy accepted responsibility for the prize, provided that certain changes were made in the will to make its provisions practicable. Regarding the prize in physiology or medicine, the Karolinska Institutet declared itself ready to assume the task of awarding it, on the condition that some changes were made in the will. As we know, the Norwegian Storting had already given its consent; the real difficulty was with the Royal Swedish Academy of Sciences. This August body had been formed in 1739 on the initiative of among others Carolus Linnaeus, and Nobel in his will had entrusted it with awarding the prizes in both physics and chemistry. Without the support of the Academy the whole project would collapse, but the Academy of Sciences dragged its feet.

The leading opponent when the Swedish Academy assumed the responsibility for the literary prize had been the eminent historian, professor Hans Forssell, and as fate would have it he was also a member of the Academy of Sciences. Having been defeated in the Swedish Academy over the literary prize, he was now firmly resolved to have his revenge when the Academy of Sciences considered the prizes for physics and chemistry. The Academy had appointed a special committee to examine Nobel's will and report on it. The committee took a favorable view of Nobel's project and in its report it recommended that the Academy should undertake to award the prizes in physics and chemistry, provided certain changes were made in the will. However, when the report was discussed in the Academy, professor Forssell succeeded in having it rejected and persuaded the Academy to table the whole question until the probate of the will had been decided in court. Seeing that this was unlikely to take place before the issue of the prize-awarding institutions had been resolved, the question of Nobel's will seemed to have reached an impasse.

To break the deadlock, the executors renewed their efforts to negotiate some kind of agreement with the Swedish Nobel relatives, who were on the verge of starting a legal action to contest the will. After much haggling a compromise was finally agreed upon, which satisfied some of the economic requests of the presumptive heirs, while at the same time it did not compromise the intentions of the will. The Academy of Sciences now reconsidered its position and agreed to undertake the task of selecting the laureates in physics and chemistry. The will of Alfred Nobel having been admitted to probate it now became possible to set up the Nobel Foundation and agree on its definitive statutes. This necessitated certain changes in the provisions of the will, which had earlier been suggested by several of the prize-awarding institutions. The following represents the most important changes and additions.

The stipulation that the prizes should be given for work done during the preceding year was obviously not practicable. Instead, it became possible to reward also older work in case its significance had not been established until recently. Furthermore, the prize could be divided between two different works and could also be awarded to several persons, who had performed a certain work together, but with the limitation that the prize could not be divided into more than three shares. The institutions awarding the prizes for physics, chemistry, physiology or medicine, and literature were to appoint Nobel Committees of three to five members to help them evaluate the candidates.

The Swedish Government approved the definitive statutes of the Nobel Foundation on 29 June 1900, while the Norwegian Storting had already in 1898 agreed upon the statutes for the awarding of the Peace Prize.

The Early Chemical Nobel Prizes

In order to be considered for a Nobel Prize, one must be nominated by somebody that the awarding institution recognizes as competent. Alternatively, the Nobel Committee in question can take its own initiative. Regarding the first prize in chemistry, invitations to submit nominations were sent to all members of the Swedish Academy of Sciences and to professors of chemistry at Swedish and certain foreign universities; all in all about 300 scientists were invited. In the period before World War I, the number of invitations increased only slowly and the nominations received were between 20 and 30 a year. Likewise the number of candidates put forward remained between ten and 20. In the period between World War I and II, there was an increase in all these numbers and after World War II the increase has been very marked. Since all Nobel laureates have the right to nominate, the growth of this group has probably played a role here.

If we look at how it all started in 1901 the scientific laureates were indeed an impressive gathering: Wilhelm Röntgen in physics, Jacobus Henricus van't

Hoff in chemistry, and Emil von Behring in physiology or medicine. These names resound like fanfares of trumpets through the history of science. To concentrate on the chemical prizes, van't Hoff had founded a new branch of chemistry — stereochemistry — but since this epoch-making work was considered too old he was instead rewarded for his work on chemical dynamics and the laws of osmotic pressure. The prize of 1902 went to one of the real giants of organic chemistry, Emil Fischer, for his work on the structure of sugars and purines. Jacobus van't Hoff was Dutch and Emil Fischer was German, but in 1903 the first Swede to be awarded a Nobel Prize, Svante Arrhenius, got the prize in chemistry for his theory of electrolytic dissociation. The first two chemical laureates had been uncontroversial and generally recognized as leaders in their fields, but about Svante Arrhenius there had been differences of opinion. Already in his thesis, he had outlined the theory of dissociation and the evaluation must have been rather discouraging. No one seemed to understand the importance of these new ideas and his thesis was awarded the undistinguished grade *non sine laude*, which the young Arrhenius obviously regarded as an insult. Perhaps professor Per Cleve in his homage to Arrhenius at the Nobel banquet hit the nail on the head when he said: "These new theories also suffered from the misfortune that nobody really knew where to place them. Chemists would not recognize them as chemistry, nor physicists as physics. They have in fact built a bridge between the two."

In 1904, Sir William Ramsay was honored for his discovery of the inert gases, like argon and helium, which meant that a whole new group of elements with similar properties were added to the periodic system. This classification of elements, suggested already in 1869 by the Russian chemist Dmitri Mendeleev, and independently by the German Lothar Meyer, was based on the observation that if the elements were arranged in order of increasing atomic weights, there was an obvious periodic recurrence in their chemical properties. Based on this regularity, Mendeleev foretold the properties and atomic weights of elements not yet discovered. When Mendeleev's predictions were later verified by the discovery of elements like gallium, scandium and germanium, his periodic system became generally accepted. Furthermore, when Ramsay's inert gases could be fitted into the system, its credibility increased even more. One might have thought that here was really a work, the full significance of which had only recently been established. This was also pointed out in a number of nominations of Mendeleev for the prize both in 1905 and 1906. So why did Mendeleev never get his Nobel Prize?

In 1905 a very strong candidate, the outstanding German organic chemist, Adolf von Baeyer, was rewarded for his work on organic dyes and hydroaromatic compounds, which had proved to be very important for the development of the chemical industry. But what of the prize for 1906, could that not have been given to Mendeleev? In fact, he came very close to getting it; the chemical Nobel

Olof Hammarsten (1841–1932).
Courtesy of the Royal Swedish Academy of Sciences, Stockholm, Sweden.

Committee had almost made up its mind to suggest his name to the Academy. However, there was one dissident who had set his heart on another candidate, the French inorganic chemist Henri Moissant, who had been nominated for his electric furnace and the isolation of fluorine. The dissident must have been very eloquent for he managed to swing the rest of the Committee behind his own candidate and Moissant got the prize of 1906.

For us with the benefit of almost a century of hindsight, it is difficult to see how an electric furnace could win over the periodic system. In any case, this was Mendeleev's last chance; he died in the beginning of 1907. It is sad that the Academy missed the possibility of adding his name to the list of chemical laureates, thereby making it even more distinguished. At the same time, it meant a somewhat unexpected opportunity for a candidate that represented an entirely different kind of chemistry.

The First Biochemical Laureate

Nowadays, biochemists are prominent in the roll of Nobel laureates both in chemistry and medicine, but it was not until 1907 that the first prize in chemistry was awarded for purely biochemical work. It is true that Eduard Buchner had been nominated already in 1905 by professor Volhard at the University of Halle, who as his first choice had Adolf von Baeyer, while Eduard Buchner was his second choice. Considering the rather low opinion that von Baeyer had of his former pupil, one wonders what he would have thought about Volhard's nomination, had he been aware of it. In any case, the Chemical Nobel Committee on this occasion decided not to evaluate Buchner (there were a number of other candidates put forward) and unanimously recommended that the prize for 1905 be awarded to Adolf von Baeyer.

In 1907 Buchner was nominated again, this time by his true friend and admirer, Carl Harries, professor at the University of Kiel, and the Swedish biochemist, Hans von Euler; in both cases with a reference to his discovery of cell-free alcoholic fermentation. There was no lack of competition, seeing that more than 20 names had been put forward, even if three of the nominees had died during the year 1907. Several of the candidates were outstanding chemists in the classical sense of the word and one takes note of names like Ernest Rutherford (nominated by Svante Arrhenius) and Walther Nernst (nominated by among others Emil Fischer). As we shall see in the following, the question of what could properly be considered as chemistry, and what ought to be seen as physiology or medicine, was brought up and led to a clash of opinions, and not for the last time. Over the years, this has remained an intensely discussed problem among the chemists in the Academy.

From the day it was founded, the Academy of Sciences has had a number of outstanding representatives of medicine among its members, including such luminaries as Linnaeus and Berzelius. In 1907, Olof Hammarsten (1841–1932) was professor of medical and physiological chemistry (as biochemistry was called in those days) at the University of Uppsala and the leading biochemist in Sweden. He was also a member of the Academy and of its Nobel Committee for Chemistry. Pictures of him later in life show a venerable old man (he lived to be 91 and continued to edit his famous *Textbook of Biochemistry* until he was 85) with a high forehead under thick white hair and a beard that made him look like the very image of a God-fearing, upright Victorian statesman. To boot he was President of Uppsala University and undeniably a figure who must have carried great weight with his colleagues in the Chemical Nobel Committee. This was the man that the Committee appointed to evaluate the merits of Eduard Buchner.

The memo that Olof Hammarsten drew up for the Committee was fairly extensive (more than ten typewritten pages) and showed him to be well acquainted with the general problem of fermentation and the long conflict between vitalists and chemists, with their respective protagonists Pasteur and Liebig. He gives every credit to Buchner's great breakthrough with the use of high pressure to disintegrate the yeast cells and produce a juice active in alcoholic fermentation, something that a number of investigators, including Pasteur, had failed to accomplish. In his memo, Hammarsten goes through the objections that had been raised against Buchner's results (remaining intact yeast cells or fragments of "living" protoplasm, etc.) and finds that Buchner has successfully repudiated these arguments. Hammarsten emphasizes that Buchner's demonstration of cell-free alcoholic fermentation once and for all puts an end to vitalism and its belief in a mystical "vital force" that did not exist outside the living cell. Furthermore, the fact that alcoholic fermentation could now be studied in a cell-free extract gave an enormous boost to what we now call enzymology, an extremely important part of biochemistry.

There is one question, though, that perhaps ought to be considered and that is the role of Buchner's collaborators. What about his brother Hans and his assistant Martin Hahn; in what way did they contribute to the great discovery? There is no doubt that Hans Buchner started the whole project of producing a cell-juice from broken yeast cells and that he later recruited his brother to the project. On the other hand, Hans wanted the juice for medicinal purposes and had no intention of looking for alcoholic fermentation. Martin Hahn, however, had a more important role directly connected with the method of breaking the yeast cells. It would seem that he was the one who originally suggested the use of high pressure in combination with kieselguhr to disintegrate the yeast. At the same time, neither he nor Hans Buchner, would seem to have paid any attention to the gas formation which resulted when sucrose was added to the juice. This crucial observation, as well as the correct conclusion that alcoholic fermentation was taking place, must be attributed to Eduard Buchner. After that, he was clearly the intellectual leader of the project. Nevertheless, one cannot help feeling that it would have been a nice gesture if Martin Hahn, and perhaps also brother Hans, would have been included as co-authors of that classical paper: "Alkoholische Gährung ohne Hefezellen."

On the basis of Hammarsten's memo, which was incorporated in their recommendation to the Academy of Sciences, the Nobel Committee for Chemistry unanimously suggested that Eduard Buchner be awarded the 1907 Nobel Prize in Chemistry and this was also the decision of the Academy. However, there were two dissenters and one of them was a man of great scientific standing in the Academy as the first and so far only Swedish Nobel laureate, Svante Arrhenius. In a letter to the Academy, he points out that his own candidate to the chemistry prize for 1907, Ernest Rutherford, had also been nominated in physics and that it sometimes becomes necessary with consultations between the Chemical and Physical Nobel Committees. He furthermore suggests that such consultations may also on certain occasions be desirable between these committees and the Medical Nobel Committee, citing the prize to Eduard Buchner as an example of this. Arrhenius is of the opinion that Buchner's work is mainly of medical interest, and to support this view he claims that Buchner's papers have appeared in medical journals. However, here Arrhenius has not done his homework properly. Buchner's pioneering reports on cell-free alcoholic fermentation in the years 1897–1902 appear in *Berichte d. Deutschen Chem. Gesellschaft*, one of the leading chemical journals at the time. Arrhenius also tries to belittle the importance of Buchner's discovery for the demise of vitalism, by pointing out, quite correctly, that Pasteur never denied the existence of what he called "organized ferments" in the cell and their importance for the vital processes. On the other hand, as we have seen previously, Pasteur clearly made a distinction between "proper fermentation," to use his own words, and the activity of soluble ferments.

Arrhenius is of course quite right when at the end of his letter, he, points out that Buchner has not really contributed anything to our understanding of the chemical processes involved in alcoholic fermentation. The punch line of his letter is as follows: "I therefore believe that Buchner's work is mainly of physiological interest and ought to have been judged by the Medical Nobel Committee." To give him his due, Arrhenius' recommendation of consultations between the Physical, Chemical and Medical Nobel Committees has indeed come to fruition and is now a matter of routine. On the other hand, there is no denying that his letter shows a complete lack of understanding of the importance of Buchner's discovery for the development of biochemistry.

Svante Arrhenius (1859–1927).
Courtesy of the Nobel Foundation, Stockholm, Sweden.

Serendipity in Research

Among the many stories told about Napoleon, most of them probably pure fabrications, there is one to the effect that before he promoted someone to a high command in his armies, he wanted to know if the man was lucky. On the whole, we have no difficulties understanding the workings of the great Emperor's mind here. He had, after all, been extremely lucky himself on many occasions, for instance at the battle of Marengo, when Desaix lost his life but saved the day for Napoleon. Maybe we must accept that being lucky is of decisive importance for generals, but surely in science it has no role to play! Nevertheless, the very word "serendipity" would seem to indicate that luck is important also in research, since the word means the ability to make fortunate discoveries when not in search of them. In fact, serendipity has come to be used with reference particularly to scientific discoveries depending on luck. This was hardly what Horace Walpole had in mind when in 1754 he wrote the play "The Three Princes of Serendip," whose unexpected discoveries created a new word in the English language.

Nowadays, the image the general public has of a scientist is probably someone, who after long consideration and often enough with a very definitive hypothesis in mind, sets out to make a number of more or less ingenious experiments to prove or disprove that hypothesis. But how often is research such an orderly process from the carefully thought out hypothesis to the actual experiment? Certainly, the answer will depend very much on what field of science we are talking about, but let us, nevertheless, try to discuss this issue in general terms.

I think most scientists would agree with me that even if you set out with certain preconceived ideas in mind, leading you to make a certain type of experiments, you often enough end up investigating a somewhat different problem. It is true that in order to get money from the Research Council or any other foundation, you have to submit a fairly detailed description of how you plan to attack the scientific problem you have outlined to them. But anyone with experience of research knows that you will in all probability have to make a number of significant changes in the project you have described. In other words, if your project represents a real piece of research, i.e. a venture into the unknown and not just a routine investigation, there is no way you could predict the outcome of it.

The German bacteriologist Robert Koch, one of the truly great names in the history of medicine, made a speech in New York towards the end of his life, where he explained his successes by saying that during his wanderings in the fields of medicine he had occasionally come to regions where gold might still be found at the wayside. Are great discoveries really something that you can stumble over, like you might find a gold nugget lying in the dust, without actually looking for it? If one considers Robert Koch himself and the kind of person he actually was, it is almost hard to believe that he really said those words. Perhaps his qualities as a scientist can be summed up in one single word: perseverance. Nevertheless, he seems to have recognized the element of chance that is inseparable from scientific research.

How often does the chance observation that leads to a great discovery occur in, for instance, biomedical research? Let us examine a few well-known cases.

The French surgeon Ambroise Paré (1510–1590) was the leading figure in surgery during the 16th century, and in his own opinion he had perfected this branch of medicine to the point when coming generations could only make relatively insignificant improvements. As a young military surgeon, he took part in the Italian war between the French King Francis I and the German Emperor Charles V. At this time, conventional surgical wisdom dictated that wounds caused by projectiles from firearms must be treated with boiling hot oil to prevent poisoning with powder from the charge that the bullet supposedly carried with it into the wound. On one occasion, when a great number of wounded soldiers were brought in for treatment, Paré ran out of oil and could not treat all his patients in the usual, barbaric way. During a sleepless night, he worried about those patients whose wounds he had only dressed but had been unable to treat properly. At the break of dawn, Paré visited them fully expecting to find them dead of powder poisoning. He was very surprised and greatly relieved when he discovered that the patients, who had not received the conventional treatment with boiling oil, were in fairly good shape compared to the ones who had been treated properly before he ran out of oil. In the latter group, all were in great pain with wound infections and fever. In his memoirs he says: "I decided never again to burn so cruelly the poor men who had wounds from gunshots." This must have been a decisive experience for Paré that greatly increased his self-confidence and helped him to become the greatest surgeon of his time. During his long career, he was remarkable for his propensity to critically examine conventional medical wisdom and to trust more in his own clinical observations.

For an older generation, the examining physician of their childhood is always visualized as a man in white tapping his fingertips on the chest of the patient and listening through the stethoscope. These diagnostic techniques, which are still in use to this day, were introduced approximately two centuries ago and represented a major improvement in our ability to examine the organs of the chest cavity. In fact, for the first time it became possible to obtain direct information about the condition of the patient's heart and lungs, something that had until then been mostly guesswork, based on the symptoms of the patient.

The Austrian physician Joseph Leopold Auenbrugger (1722–1809) was the son of a wine merchant and according to tradition, he had as a child observed how his father tried to judge the level of wine in the barrels by tapping on the wall of the cask and listening to the quality of the resulting sound. Maybe it was the memory from his childhood of his father examining the half-full wine casks that gave Auenbrugger the idea of trying to judge the condition of the patient's lungs by tapping on the chest wall, using his fingers as a kind of hammer. He then discovered that the quality of the sound was very different over the normal, air-filled lung, which gave a drum-like sound, compared to the dull sound that resulted when, for instance, the pleural sack was filled with liquid. Auenbrugger used this method, which came to be called percussion, also to determine the size of the heart or the amount of liquid that in certain conditions is found in the heart sack. Of particular importance was the fact that he systematically tried to correlate his percussion findings to the observations at the autopsy. As has so often been the case with major breakthroughs in medicine, the conservatism of the profession was a major obstacle and it was not until a few years before Auenbrugger's death that the great French clinician Jean Corvisart, revived percussion and demonstrated its great importance, at the same time giving Auenbrugger full credit for his discovery.

René Laennec (1781–1826) had studied medicine in Paris, which brought him in contact with some of the leading figures in French medicine, such as Corvisart and Xavier Bichat, the great pathologist, whose collaborator Laennec became. He was particularly interested in diseases of the chest and when he was appointed physician-in-chief at the l'Hopital Necker in Paris, he began to listen to the sounds coming from the patient's chest cavity by laying his ear to the chest wall. This method of auscultation, as it was called, had been known since the days of Hippocrates, but Laennec introduced an important technical improvement. It came about mainly by chance in the following way. One day, he was trying to examine a young female patient with a suspected heart condition. Laennec wanted to listen to the heart sounds but the patient was so obese that it was virtually impossible to use the time-honored method with the ear against the chest. He then recalled that as a child he had amused himself by listening with his ear against one end of a wooden beam while a playmate scratched with a needle at the other end. Laennec particularly remembered how distinct he had perceived the weak sounds of the needle through the beam. He took a sheet of paper and rolled it as tightly as possible so that it formed a short staff. Placing one end of the staff against the patient's chest and listening at the other end, he was surprised by how clearly he heard the heart sounds through the mighty fat layers of his patient. He immediately realized the great usefulness of this simple instrument. Unlike Auenbrugger's percussion, Laennec's improvement of auscultation, the stethoscope, was an instant

René Theophile Laennec (1781–1826).
Courtesy of Bonnierförlagen, Stockholm, Sweden.

success and to this day it remains the symbol of a doctor. The stethoscope made Laennec famous, he became a full member of the Académie de Medicine and was awarded the Legion of Honor. Unfortunately, he was not able to enjoy his celebrity for very long. At the age of only 45 he died of tuberculosis, a disease that had claimed several victims in his family, including his mother.

The cases we have examined so far all demonstrate that an element of chance is sometimes involved in scientific discoveries. But is it really correct to call it serendipity? The three princes of Serendip in Horace Walpole's play certainly made fortuitous discoveries, but more than that, they were in no way looking for what they eventually found. On the contrary, they were pursuing entirely different goals and their successes bear no relation to what they originally set out to do. The difference between the cases discussed above and the heroes of Walpole's play is obvious. Paré, Auenbrugger and Laennec were one way or another favored by chance, but on the other hand they were clearly engaged in activities that were directly related to their discoveries. Could one possibly think of another case, even more similar to the princes of Serendip?

To us it is obvious that every effort should be made to eliminate or at least reduce the pain of any surgical or dental operation. However, successful efforts to achieve anesthesia only began towards the middle of the 19th century and the real pioneers here were the dentists. One of them was the young American dentist Horace Wells. He had gone to the marketplace to amuse himself so as to forget a toothache that he was suffering from. At the market was an itinerant quack that administered laughing gas (nitrous oxide) to anyone who wanted to test it. His customers became intoxicated and their antics were so violent that they sometimes hurt themselves. Horace Wells noticed that on these occasions the victims did not seem to feel any pain, even when the injuries were fairly extensive. He was struck by the idea that maybe laughing gas could be used to reduce the pain in dentistry and why not try it in his own case? With the help of the quack and a dentist colleague, Horace's aching tooth was successfully extracted with minimal pain, and after the operation Wells exclaimed: "A new era in tooth-pulling!"

He began to use laughing gas in his practice with great success and everything was fine until he managed to persuade a well-known surgeon in Boston to use his laughing gas at an operation. However, something went wrong. Perhaps an insufficient amount of gas had been administered, the patient cried out during the operation, everyone present laughed at Wells (with the exception of the patient presumably) and the poor pioneer of anesthesia was devastated. In his despair he took to abusing the new anesthetics, ether and chloroform, that had been introduced by others after his own experiment with laughing gas. In the end, he committed suicide while intoxicated with chloroform.

Horace Wells (1815–1848).
Courtesy of Bonnierförlagen, Stockholm, Sweden.

A sad story, indeed, and perhaps there is more serendipity here than in the previous examples. Nevertheless, in my opinion nothing can compare with Buchner's discovery of cell-free alcoholic fermentation where serendipity is concerned. It really has everything. There is his brother Hans, who wanted to use the press-juice from yeast for medicinal purposes and dumped in sucrose to preserve the juice without noticing the formation of gas, or at least without drawing any conclusions from it. He had never worked with alcoholic fermentation, nor had his brother Eduard, with the exception of a single publication more than ten years earlier. However, when Eduard arrived from Tübingen and happened to be present at the conservation of the juice with sucrose, he immediately drew the correct conclusion. The salient point is that whatever prompted him to participate in the project certainly had nothing to do with an interest in alcoholic fermentation. He was there just to help his brother Hans whom he admired and felt greatly indebted to. The great discovery came about without any intention on the part of the two brothers or their assistant, Martin Hahn. It is a case of pure serendipity.

Bibliography

Berthelot, Marcelin. "Sur la Fermentation Glucosique du Sucre de Canne." *Comp. Rend.* **50**, 980–984 (1860).

Buchner, Eduard. "Alkoholische Gährung ohne Hefezellen." *Ber. Chem. Ges.* **30**, 117–124 (1897).

Buchner, Eduard; Buchner, Hans; and Hahn, Martin. *Die Zymasegährung* (Oldenbourg, Munich, 1903).

Cagniard–Latour, Charles. "Mémoire sur la Fermentation Vineuse." *Ann. Chim. 2e Sér.* **68**, 206–223 (1838).

Dubos, René. *Louis Pasteur: Free Lance of Science* (Little Brown, Boston, 1950).

Fischer, Emil. "Bedeutung der Stereochemie für die Physiologie." *Z. Physiol. Chem.* **26**, 60–87 (1898).

Fruton, Joseph S. *Molecules and Life* (Wiley–Interscience, New York, 1972).

Geerk, Frank. *Paracelsus, Arzt unserer Zeit* (Benziger Verlag, Zürich, 1992).

Harries, Carl. "Eduard Buchner." *Ber. Chem. Ges.* **50**, 1843–1876 (1917).

Jorpes, Erik. *Jacob Berzelius: His Life and Work* (Almquist and Wikzell, 1970).

Kathe, Johannes. *Robert Koch und sein Werk* (Akademie Verlag, Berlin, 1961).

Kühne, Friedrich Wilhelm. "Erfahrungen und Bemerkungen über Enzyme und Fermente." *Untersuchungen aus dem Physiologischen Institut Heidelberg* **1**, 291–324 (1878).

Kützing, Friedrich T. "Mikroskopische Untersuchungen über die Hefe und Essigmutter." *J. Prakt. Chem.* **11**, 385–409 (1837).

Liebig, Justus von. "Über die Gährung und die Quelle der Muskelkraft." *Ann. Chem.* **153**, 1–47; 137–228 (1870).

Lindeboom, Gerrit Arie. *Hermann Boerhaave* (Methuen, London, 1968).

McKie, Douglas. *Antoine Lavoisier* (Constable, London, 1952).

Melhado, Evan M. and Frängsmyr, Tore (eds.) *Enlightenment Science in the Romantic Era* (Cambridge University Press, 1992).

Möllers, Bernhard. *Robert Koch. Persönlichkeit und Lebenswerk* (Schmal und von Seefeld Nachf., Hannover, 1950).

Partington, James Riddick. "Jeremias Benjamin Richter and the Law of Reciprocal Proportions." *Ann. Sci.* **7**, 173–198 (1951).

Partington, James Riddick. "The Life and Work of John Mayow." *Isis* **47**, 217–230; 405–417 (1956).

Partington, James Riddick. *History of Chemistry* (MacMillan, London, 1961–1970).

Pasteur, Louis. "Mémoir sur la Fermentation Alcoolique." *Ann. Chim. 3e Sér.* **58**, 323–426 (1860).

Patterson, Elizabeth C. *John Dalton and the Atomic Theory* (Doubleday Co., New York, 1970).

Ramsay, William. *The Life and Letters of Joseph Black* (Constable, London, 1918).

Read, John. *Humour and Humanism in Chemistry* (London, 1947).

Schück, Henrik; Sohlman, Ragnar; Österling, Anders; Liljestrand, Göran; Westgren, Arne; Siegbahn, Kai; Siegbahn, Manne; Schou, August; and Ståhle, Nils. *Nobel, The Man and His Prizes* (American Elsevier Publishing Co., New York, 1972).

Schwann, Theodor. "Vorläufige Mitteilung, Betreffend Versuche über die Weingährung und Fäulnis." *Ann. Phys.* **41**, 184–193 (1837).

Taylor, Frank Sherwood. *The Alchemists, Founders of Modern Chemistry* (Henry Schuman, New York, 1949).

Underwood, Ashworth. "Franciscus Sylvius and His Iatrochemical School." *Endeavour* **31**, 73–76 (1972).

Willstätter, Richard. *Aus Meinem Leben* (Verlag Chemie, Weinheim, 1949).

Index

Date Due

MAR 2? 2006			
DEC 1 1 2007			
DEC 0 6 2007			